ARID SHRUBLAND PLANTS

OF
WESTERN AUSTRALIA

ARID SHRUBLAND PLANTS

OF
WESTERN AUSTRALIA

A. A. Mitchell
&
D. G. Wilcox

with photographs by E. Laidlaw

SECOND
AND ENLARGED EDITION

UNIVERSITY OF WESTERN AUSTRALIA PRESS
IN ASSOCIATION WITH THE

DEPARTMENT OF AGRICULTURE
WESTERN AUSTRALIA

First published in 1988 by the
University of Western Australia Press
Nedlands, Western Australia 6009
with the Department of Agriculture, Western Australia

This revised and enlarged edition first published in 1994

This book is copyright. Apart from any fair dealing for the purpose of private study, research, criticism or review, as permitted under the Copyright Act, no part may be reproduced by any process without written permission. Enquiries should be made to the publisher.

© Department of Agriculture, Western Australia, 1988, 1994

National Library of Australia
Cataloguing-in-Publication data

Mitchell, A.A.
  Arid shrubland plants of Western Australia.

  2nd. and enl. ed.
  Includes index.
  ISBN 1 875560 22 X PB.
  ISBN 1 875560 47 X HB.
  1. Arid regions plants — Western Australia. 2. Shrubland ecology — Western Australia.
  3. Range management — Western Australia. I. Wilcox, D.G. II. Laidlaw, E. (Ernest),
  1946–   . III. Western Australia Dept. of Agriculture. IV. Title.

581.9941

Consultant editor Suzanna Norton
Design by Susan Ellvey, Designpoint, Perth
Typeset in 11 point Adobe Garamond by Lasertype, Perth
Colour separations and negative preparation by Fonts Plus, Perth
Printed by Scott Four Colour, Perth

CONTENTS

Foreword to the first edition	vii
Foreword to the second edition	viii
Acknowledgements	ix
The arid shrublands of Western Australia	1
How to use this book	2
The aim of rangeland management	5
Managing rangeland resources	6
The plants	6
Soil	9
Setting range management objectives	12
Setting the stocking rate	12
Pasture decisions	12
Rehabilitation techniques	15
Mine site rehabilitation in the arid shrublands	15
Arid land rehabilitation	21
Plants in this book	25
Section I Annual herbs, short-lived perennials, ferns and vines	25
Section II Annual and perennial grasses	99
Section III Shrubs less than 3 m high	155
Section IV Trees and shrubs more than 3 m high	365
The origin of the generic names	456
Indexes	
Scientific names and authorities	461
Common names	471

FOREWORD TO THE FIRST EDITION

Pastoralists in the shrublands of Western Australia are constantly making judgements affecting the well-being of the almost bewildering variety of native plants which make up the rangelands. Each decision on stocking rates, season of use, or whether some country should be stocked at all, influences the persistence of seedlings, seeding efficiency, the build-up of drought reserves, root development and the vigour of adult plants.

The stocking policies adopted affect the composition of the rangeland. Some plants are favoured and others are disadvantaged by the managers' decisions. On the one hand a desirable mix of plants giving the most consistent production will be maintained. On the other hand the vegetation may be driven so hard that it consists of only undesirable or ephemeral plants of little grazing value.

Clearly, then, rangeland managers need to understand the complexities of the resource they use and to be able to distinguish desirable from undesirable plants in each type of rangeland. They should be able to identify these plants with confidence and be able to define rangeland management goals in terms of the number and species of plants wanted on the property.

This book is a development of an earlier publication of the Department of Agriculture in which David Wilcox and John Morrissey described some of the important rangeland plants of the shrublands.

Andrew Mitchell and David Wilcox, the authors of this book, have produced descriptions of many more of the important species. They have described the value of the plants for grazing purposes and have classified them on the basis of their usefulness for indicating rangeland condition; the vital characteristics used in determining grazing policies.

The pastoral shrubland areas of Western Australia are a valuable resource peculiar to this State. We, the present users of this resource, should remember that it is not ours to squander. We should rather regard ourselves as its custodians for future generations.

This book provides practical advice on the management of plant resources so that we can properly discharge our given responsibilities. The authors have had a lifetime of experience in the shrublands. They have set down the results of patient observation, scientific research and the collective wisdom of pastoralists, so that the rangeland user will be able to make decisions ensuring the well-being of the industry and the maintenance of our common heritage.

The photographs taken by Ernie Laidlaw of the Department of Agriculture are a tribute to his professional skill. They will ensure easy identification. But they are more than this: they have been composed with an artistry and sympathy, giving the subjects a quality which will encourage the use of this book.

N.J. Halse
Director of Agriculture

FOREWORD TO THE SECOND EDITION

The first edition of this book was immediately popular with pastoralists in the first instance but also with others who use the pastoral zone, including mining companies and geologists, tourists and those who enjoy travelling and camping in the outback. Within 18 months of its publication, stocks of the first edition were exhausted and it quickly became obvious that a second printing would be well received.

This edition has been made possible with the generous support of several mining companies with interests in the arid shrublands. It has also been supported by a grant from the National Soil Conservation Program, as information on the role of the important pasture plants is central for the conservation of the natural resources of the land.

The authors have added 83 species to those in the first edition and have included a number of plants which have special roles in mine site rehabilitation. They have enlarged the text to include comments on the place of these plants in reclamation work, and have added sections on techniques for the re-establishment of vegetation on disturbed land, waste dumps and degraded grazing land.

The photographs of the plants added to this edition have again been taken by Ernie Laidlaw. He has once more demonstrated his ability to convey subtlety of form, colour and tone in works of great artistic merit, which will make this book the standard work of professionals and lovers of the interior of Western Australia.

M.D. Carroll
Director General of Agriculture

ACKNOWLEDGEMENTS
DEPARTMENT OF AGRICULTURE

The publication of this book was made possible through the generous support of the mining companies listed below.

Arimco Mining Pty Ltd	Metana Minerals NL
Asarco Australia Ltd	Mt Gibson Gold Project
Ashton Gold — Laverton Operations	Mt Martin Gold Mines NL
Australian Consolidated Minerals	Newcrest Mining Limited
Big Bell Mines Pty Ltd	Normandy Poseidon Limited
Central Kalgoorlie Gold Mines NL	Pancontinental Goldmining Areas Pty Ltd
Coolgardie Gold NL	Pancontinental Gold Operations
Dominion Mining Limited	Paragon Resources NL
Eastmet Limited	Peak Hill Resources
Goldfan Limited	Placer (Granny Smith) Pty Ltd
Gwalia Group	Plutonic Operations Limited
Hill 50 Gold Mine NL	Plutonic Operations Ltd, Lawlers
Homestake Australia Limited	St Barbara Mines Ltd
Posgold — GMK Jubilee Mine	WMC Kalgoorlie Nickel Smelter
Karonie Gold Project	WMC Leinster Nickel Operations
Metall Mining Australia Pty Ltd	WMC Windarra Nickel Project

A grant from the National Soil Conservation Program funded by the Government of the Commonwealth of Australia also substantially assisted the production.

ACKNOWLEDGEMENTS

AUTHORS

The authors are indebted to the Western Australian Department of Agriculture for substantial support in the publication of this book. Our gratitude is expressed to the Director General, Dr Mike Carroll, for this support, and to his staff for their contribution. We particularly recognise the assistance of the Information and Media Services Branch including Ernest Laidlaw, photography, Rod Lewis, initial layout concept, and Bev Croucher and Yvonne Vogel, word processing.

Brett Ward from the Katanning District Office contributed the text on rehabilitation of arid land. John Stretch from the Carnarvon District Office provided the texts for *Acacia ligulata, A. subtesseragona* and *Spartothamnella teucriiflora*. Philip Thomas from the Land Management Branch, South Perth, prepared the plant distribution maps. Pam Booker and Shelly Ford of the Document Support Centre produced the final document for typesetting. The publication was managed by Alec Holm from the Land Management Branch, South Perth, with editorial advice from Glenis Ayling.

The staff of the Herbarium of Western Australia were unfailingly helpful in verifying the plant identifications.

Mary Fletcher, Karratha, and Barbara Jennings of Kalgoorlie Consolidated Gold Mines Pty Ltd, Kalgoorlie, assisted one of the authors, D.G. Wilcox, with the text on rehabilitation of mine sites.

Specialist botanical identifications were provided by Dr G. Benl (Munich), Dr R.J. Chinnock (Adelaide), R.J. Cranfield (Perth), Dr M. Lazarides (Canberra), Dr B. Maslin (Perth), and Dr Paul G. Wilson (Perth).

References consulted during the preparation of some descriptions included:

- *Flora of South Australia.* Black, J.S. 1948.
- *Plants of Western New South Wales.* Cunningham, G.M., Mulham, W.E., Milthorpe, P.L., and Leigh, J.H. 1981.
- *Flora of Central Australia.* ed. Jessop, J. 1981.
- *The Grasses of Central Australia.* Lazarides, M. 1970.
- *Flora of Australia,* various volumes.

We have also been kindly allowed to base some of our descriptions on as yet unpublished manuscript descriptions by Dr J. Thompson of the New South Wales Herbarium and Dr R.J. Chinnock of the South Australian Herbarium.

A.A. Mitchell and D.G. Wilcox

The arid shrublands of Western Australia

The arid shrublands pastoral region covers about 600,000 sq km of semi-arid to arid shrublands where rainfall is insufficient to grow crops with confidence.

The Western Australian wheatbelt forms the south-western boundary which follows the 275–250 mm rainfall isohyet. The south-eastern boundary is provided by the Great Victoria and Gibson Deserts where rainfall is only sufficient to support sparse shrubs and spinifex. The north-eastern boundary lies along the Fortescue River in an area characterised by a lack of reliable winter rain.

The climate is characterised by hot summers, with temperatures up to 47°C, and very low humidity. Summer rains are more common in the north than the south. In the south, the winters are cold with regular frosts, and maximum temperatures of 10°C are not uncommon. Temperatures gradually increase to the north. Winters in the northern shrublands are, therefore, short and mild. Winter rains are reasonably reliable and usually start in May or June, but consecutive years of drought, when little effective rain falls, occur frequently.

The shrublands support about 320 pastoral leases or stations, which currently carry about 2.2 million sheep and 80,000 cattle. The average size of a lease is about 250,000 ha. The animals graze the native vegetation as there has been no pasture improvement. The southern Pilbara and east Gascoyne stations run most of the cattle whilst sheep are run throughout the remainder.

Stocking rates are low and vary from about one sheep to 5 ha to one sheep to 40 ha. Sheep stations are usually completely fenced into paddocks of about 6000 ha in area. Most cattle stations are unfenced, but may have one or two holding paddocks. The land has few permanent water holes, and most stock water is pumped from bores or wells using windmills. Earth dams are used to collect stock water in the Goldfields and the western Gascoyne where ground water is either very deep or saline.

The area is sparsely populated. The coastal towns are based on a number of industries, including tourism, but the inland towns now depend on mining for their prosperity.

How to use this book

The plants in this book are arranged in four sections according to their height and growth form:

Section I   Annual herbs, short-lived perennials, ferns and vines

Section II   Annual and perennial grasses

Section III Shrubs less than 3 m high

Section IV   Trees and shrubs more than 3 m high

The section title pages highlight the growth forms and relative heights of the plants in their sections, as illustrated below for Section III.

The purple shading on the maps indicates the distribution of each species in Western Australia. No attempt has been made to extend the information on distribution to other States. The regions referred to in the text are shown on the map on page 4.

Where available, estimates of the principal indicators of forage value (crude protein, digestibility and salt content) have been supplied. In general terms, forage with less than 8 per cent crude protein and less than 50 per cent digestibility is insufficient for herbivore maintenance.

High salt content will limit the usefulness of some plants. The salt contents of plants in this category are given.

Section III
Shrubs less than
3 m high

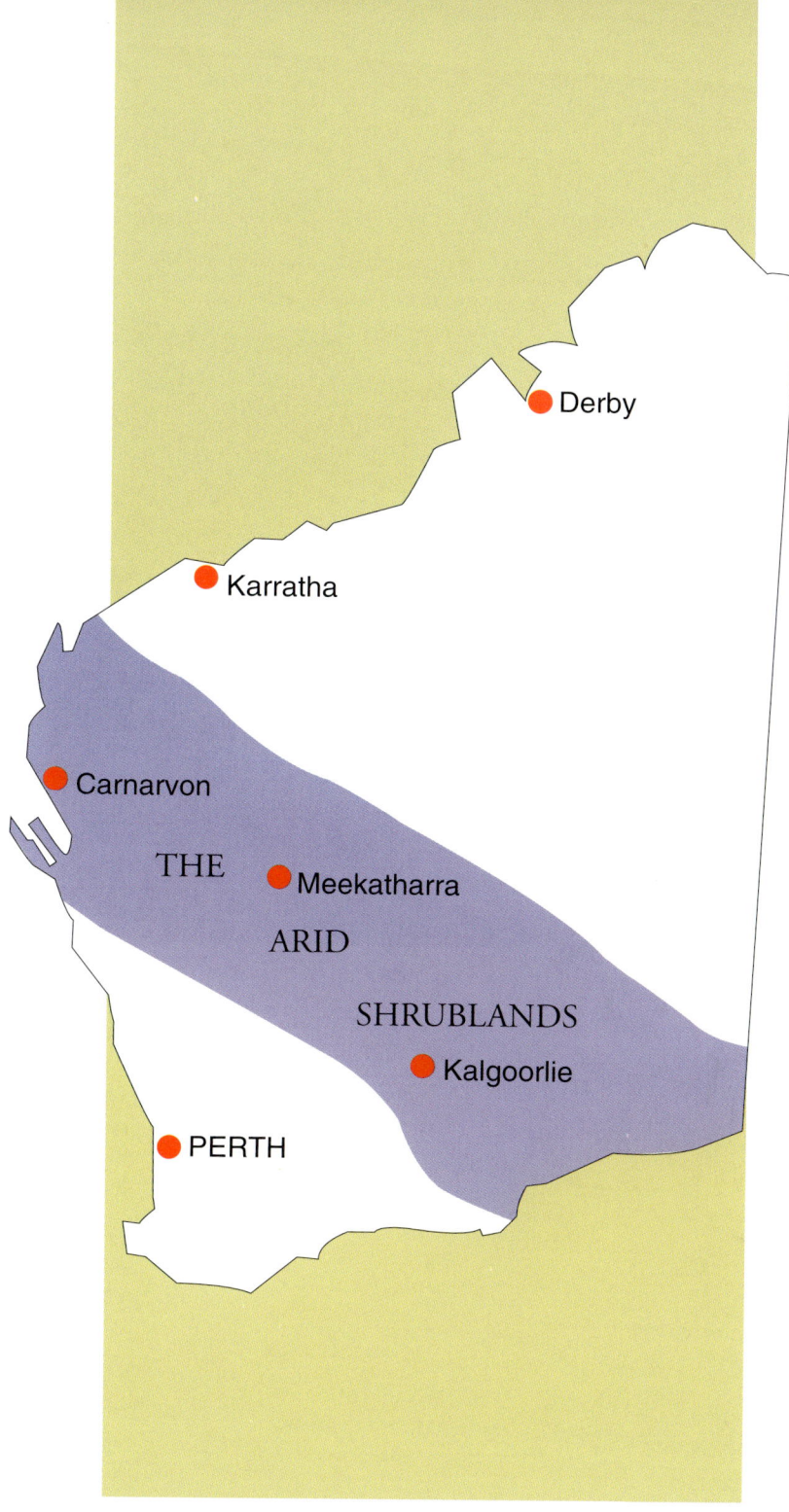

SPECIES DISTRIBUTION — throughout this book the areas shown in purple indicate the distribution of each species, as in the above example.

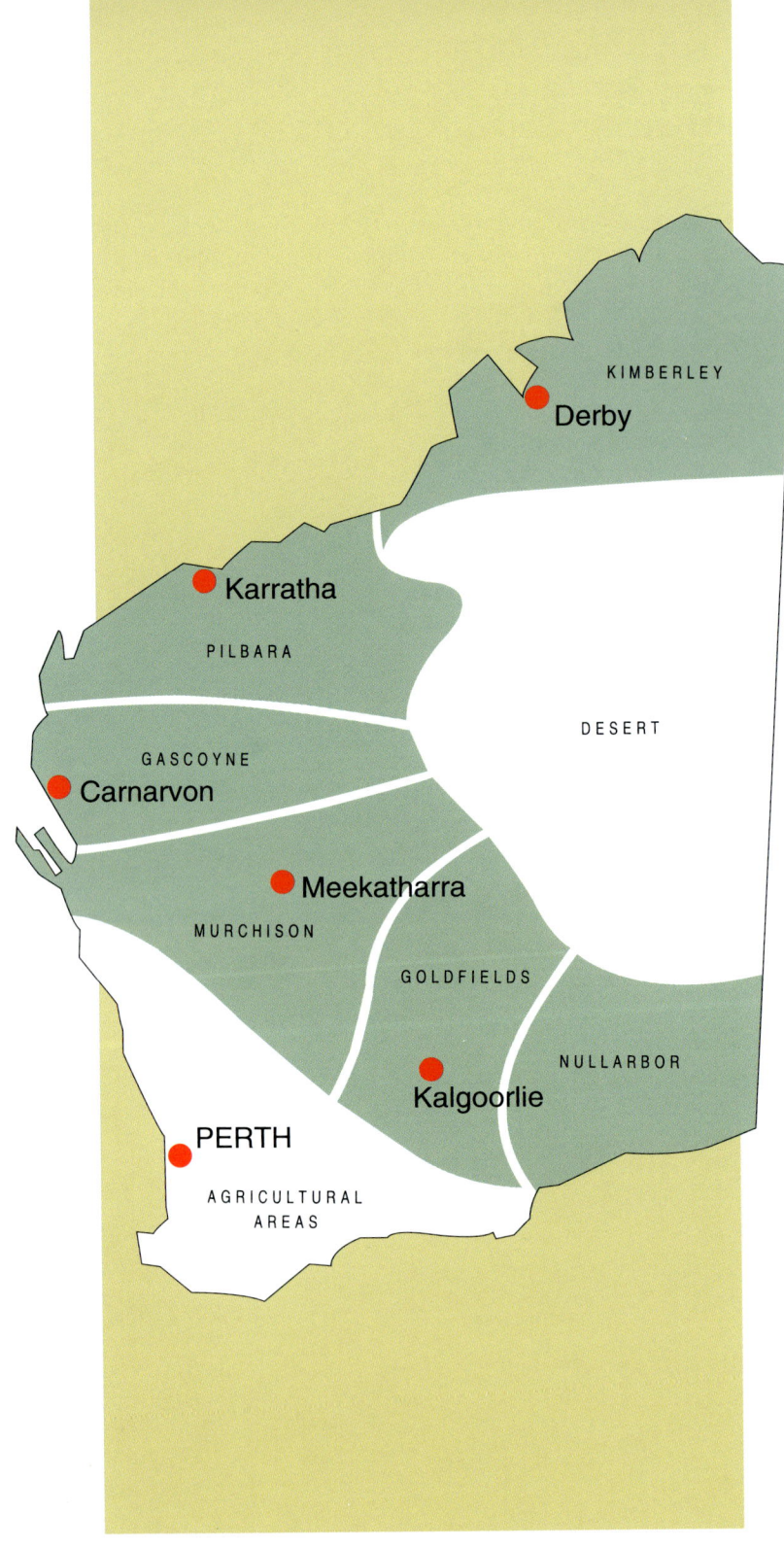

REGIONS REFERRED TO IN THIS BOOK

The aim of rangeland management

The aim of rangeland management is to maintain the basic resource of soil and plants in a stable and productive condition. The condition or 'health' of these resources determines the number of stock which can be carried on a property. The sound management of rangeland pastures demands some understanding of the complex ecological processes involved in grazing in a low rainfall environment.

Long-term productivity is achievable when rangelands:
- are covered with the most productive and resilient plants possible;
- are covered with the most palatable plants possible;
- are stable; and
- can absorb rainfall easily.

Poor management practices lead to rangelands which:
- are unproductive;
- are dominated by unpalatable plants;
- are unstable; and
- shed rainfall instead of absorbing it.

Sheep in a good winter season at Carnarvon

Managing rangeland resources

THE PLANTS

The plants of the arid shrublands can be classified on the basis of their life forms. They include:

- Annual grasses and herbs such as red grass, crowsfoot and everlastings, which produce a spectacular flush of growth in good seasons, but are absent in poor seasons.
- Annuals which complete their whole life cycle in one, short season. The quality of the forage they provide deteriorates rapidly once they mature. When dry, these plants lose their anchorage to the soil and are broken down by wind, termites, ants and micro-organisms.
- Weakly perennial plants, such as wind grass and bindiis, which behave like perennials through a continuous series of good seasons, but die in adverse seasons.
- Perennial grasses such as broad-leaved wanderrie grass, soft wanderrie grass and spinifex, which become dormant during dry periods and spring to life again after rain, provided that the dry period has not been too long. These plants can grow on rainfall amounts inadequate for annual grasses and herbs.

Cattle in mulga country in the southern Pilbara

A fenceline effect in the Murchison. The paddock, on the left, has been severely overgrazed

- Trees and shrubs which persist in the dry years. Some, such as mulga, turpentine bush and Wilcox bush, retain their leaves in the driest years, while others, such as flannel bush and cotton bush, drop their leaves once the soil becomes dry.

The different classes of rangelands that we recognise, such as wanderrie or saltbush country, carry different combinations of these basic life forms, though some species such as wind grass occur in all rangelands.

Plant Competition and Survival Under Grazing

In the low rainfall environment of the shrublands, plants are in constant competition with one another for water and nutrients. Grazing by domestic stock has upset the balance developed under natural conditions and, in some cases, adversely affected the growth, survival and establishment of some of the valuable components of the pastures. As a consequence, less valuable species have prospered at their expense and then spread to become the dominant plants in what were productive lands.

These less valuable and, therefore, undesirable species may be unpalatable and, as a result, reduce the amount of feed available. Alternatively, they may be drought-evading, dropping their leaves in drought, or ephemeral and unable to respond to the small amounts of rain which characterise the drought years.

Some plants may be undesirable in one pasture type and yet acceptable in another. Wind grass is undesirable in wanderrie pastures, but it is valuable on the stony, hill pastures where it has a much higher protein content

The degree to which the valuable species are displaced in a pasture by less useful species depends upon the resilience of the plants concerned. Some species, such as sago bush and Gascoyne bluebush, are particularly resistant to grazing, while others, such as warty fuchsia bush and shy bluebush, are susceptible to grazing and, as preferred species, may scarcely tolerate it at all, even under recommended stocking rates for the pasture type.

While palatability affects the survival rate of a plant species in any one community, the concept of preference must also be taken into account if the same species occurs in a range of communities. In some communities a plant will have a high preference status, but in others a lower status. For instance, Wilcox bush in some mulga communities is a preferred plant, but not on the coastal plain in the Gascoyne and on gravel hills as more palatable species are available.

Indicator Value, Decreaser and Increaser Species

The foregoing short outline has shown that the art of range management lies in the ability of the rangeland user to maintain a balanced and stable pasture, consisting of a mix of plants of differing palatability and resilience. Even though there are problems created by variations in palatability and preferences in different rangeland types, it is possible to place some plants in one of two categories, those of decreaser or increaser. Plants which can be so designated have what is termed 'indicator value'.

Desirable plants — those having the advantages of palatability, drought resistance and resilience but which disappear from a rangeland when it is abused — are known as 'decreasers'. As the number of decreaser species diminishes, range condition declines.

Less desirable plants and even undesirable plants which replace the decreaser species are known as 'increasers', since they occupy the places recently vacated by the decreasers when management was at fault. As suggested, not all increaser species are valueless, but they are usually less productive than the species they replace. Some increasers, such as needlebush or sandbank poverty bush, are worthless and when dominating a rangeland type, make it unproductive.

Decreaser and increaser species are said to have 'indicator value' since their presence in, or absence from, a rangeland type *indicates* the health of that particular rangeland and its capacity to support grazing animals. Needlebush or mulga bluebush are good indicators of range condition. Some plants have no indicator value since they do not respond at all rapidly to changes in the health of the rangeland. Plants in this category include mulga and sago bush. Others may only reflect current seasonal conditions.

The value of the plants in this book as indicators is included in the descriptions where relevant. As already noted, when dealing with the concept of desirable plants, some plants may be decreasers on one rangeland type, and yet increasers in another. Before assessments of range condition are made using decreasers and increasers, it is important that the status of the plant as an indicator be known for that particular rangeland. Where necessary these observations are made in the text.

Assessing Plant Condition

When rangeland condition is assessed using indicator plants, it is important that as many species as possible are used. Misleading interpretations may result if only one or two species are used.

If decreaser species, or those with little indicator value, make the major contribution to forage at a site, the rangeland can be described as being in good condition. Conversely, if increaser species of low palatability are the dominant plants on a site, then the rangeland is in poor condition.

SOIL

In the arid shrublands the principal factor affecting production is the amount of water stored in the soil after rain, though nutrient status also plays a significant role.

Any factor which reduces the capacity of a soil to absorb rainfall lessens the amount of forage which can be produced. Reduced infiltration and the loss of valuable rainfall to run-off are a natural consequence of soil compaction and soil sealing caused by overgrazing. Wind erosion and water erosion, themselves the product of overuse, also reduce the amount of rainfall absorbed by the soil.

Correct soil management will include the retention of as much plant litter as possible on the surface and the maintenance of overall plant cover to protect the soil surface.

Severe erosion around a snakewood tree in the south-west Pilbara

As with the plant resources, the condition or 'health' of the soil resources may also be assessed. The soil is in good or excellent condition if:

- it is free of gutters;
- there are no scalds;
- there are no large hummocks of soil beneath trees and shrubs;
- the soil surface is flaky or crusted;
- the surface is littered with dead plant material.

The assessments made of the plant and the soil condition are usually combined to give an estimate of the range condition of that pasture.

Good range condition — shrub cover is good and soils are intact

Fair range condition — shrub cover is patchy and footslopes are eroding

Poor range condition — shrubs are eliminated and the whole soil surface is eroding

Setting range management objectives

Range management is a planned response of the land manager to seasonal conditions or rainfall and to the health of the rangeland. Thus, an effective manager actively collaborates with nature to maintain the rangeland in good condition, while at the same time meeting the feed requirements of stock and budgetary demands.

The manager continually has to make decisions on stock numbers and season of use, in the context of a need for an adequate income and of the needs of the pastures used to obtain this income.

SETTING THE STOCKING RATE

Stocking rate is the most important decision made by the station manager. The number of stock that can be carried on a property is influenced by range type, range condition, paddock size, the distribution of the range types within a paddock, the number and spacing of the stock water points, the quality of the water supplies, and the previous stocking history of the property.

Each station has its own unique combination of these factors. There is, therefore, no simple formula which can be used to make a decision on the number of animals to be carried. However, the following guidelines are useful:

- The number of stock should be set so that the range condition of the property is maintained or gradually improved. Obviously, the stocking rate required to obtain this improvement on a property in poor range condition would be lower than that for a similar property in good condition.
- The number of stock carried on a property should allow some areas of the property to be free of stock, as described below.
- The long-term average stock-carrying capability after providing areas of rest is a useful guide to the stocking level which can be adopted in the first instance.
- The temptation to stock the property heavily in good seasons should be resisted. As suggested earlier, feed not consumed by grazing animals is not wasted.
- The table shows the variation in dry matter production that can occur in various rangeland types. Long-term stocking rates for these are suggested on the assumptions that a grown sheep requires about 400 kg of dry matter per year and that between 20 and 30 per cent of the feed produced is eaten.

PASTURE DECISIONS

Pasture decisions should be made in a framework of long-term objectives, which will span decades and which will turn on the needs of the pastures for maintenance and regeneration. The pasture objectives to be set by the manager will depend upon the condition and the type of pasture as well as upon the rainfall.

Pastures in good condition may benefit from periodic rests to allow the build-up of energy reserves, new plants to replace those that have died and seed set. Usually a spell over a winter season is sufficient. A rest from grazing in a poor season is not usually as

Dry matter production and suggested stocking rates of various rangeland types

Rangeland type	Production kg/ha		Suggested long-term stocking rate
	Best season	Worst season	ha/sheep
Saltbush (Yalgoo)	560	40	8
Mulga–wind grass herbage (Wiluna)	825	Nil	24
Mulga, stony plains (Meekatharra)	130	11	20
Wanderrie (Meekatharra)	110	16	16
Bluebush, stony plains (Gascoyne)	350	40	16
Saltbush (Nullarbor)	1000	100	8

effective as one in a good season but it has been found that new plants can be added, even under adverse conditions.

Each class of rangeland makes different management demands. In saltbush country, if soil erosion is not active and sufficient parent plants remain to provide seed, short-term rests of about six months may help to maintain productivity and vigour in the stand. On the other hand, if the rangeland is in such poor condition that soil erosion is active and most of the desirable species are absent, it will be necessary to remove stock for longer periods. Following the rest from grazing, the stocking rate should be adjusted so that the plant cover re-established during the rest period is maintained.

Wanderrie pastures normally do not require the same length of spelling as saltbush does. Complete closure of wanderrie pastures for two good summer seasons can repair them even when they are in poor condition. In less severe cases, one good summer season should be sufficient.

Degraded mulga shrublands and stony hill pastures should receive a year of rest in good seasons, provided that soil erosion is not active. In cases where soil erosion is a problem, the rest should be extended until some improvement is obvious. In some circumstances, it may be necessary to control concentrated water erosion with contour banks or cultivated strips. In some instances, short rests over the winter or summer growing season can be beneficial.

The amount of rainfall received will influence the length of the rest required. In

The result of severe overgrazing by sheep and feral goats

The result of two years' spelling from grazing

high rainfall years, gains from resting can be spectacular. Conversely, regeneration after resting in low rainfall years, or in years when follow-up rains are inadequate, may be less obvious.

Rainfall in the colder months is usually ineffective in promoting the germination of perennial grasses and shrubs. Therefore, pastures containing these species will respond only slowly to rest following these rainfall events. However, the vigour of the existing plants will be enhanced.

Rehabilitation techniques

MINE SITE REHABILITATION IN THE ARID SHRUBLANDS

Mining operations are scattered throughout the arid shrublands of Western Australia. These mines generally operate within pastoral leases and often in areas which have been disturbed by mining in the past. While the actual phase of active mining is relatively short-term, the impact of mining on the native flora and fauna can be locally significant and long-term in effect.

Public perception and expectations of the mining industry have changed dramatically in recent years. There are now rehabilitation standards set by the Minerals and Energy Department with which mining companies are legally bound to comply, regardless of the size or type of mine in operation. With adequate planning and a comprehensive understanding of the requirements, these standards can be met at the lowest cost if addressed throughout the active stage of the mine life.

This chapter addresses the role of vegetation in the rehabilitation process.

Planning for Rehabilitation

Key themes in environmental planning should be minimal disturbance of the environment and an appropriate budget allocation in the planning stage.

To the extent practicable, rehabilitation should be undertaken progressively during the mine life.

There are three main elements that should always be included in the rehabilitation plan:

- Rehabilitation objectives
- A description of the site
- A detailed plan of the site.

REHABILITATION OBJECTIVES

Clear rehabilitation objectives consistent with the projected future land use of the area must be defined.

Long-term rehabilitation objectives may vary considerably at different sites. In all cases the first objective will be to protect the safety and health of all people living in the area.

The final objective of rehabilitation is usually reclamation of the site to a safe, stable and non-erodible condition. Engineering design has the greatest impact on basic stability and on the visual impact of the by-products of mining. The role of vegetation is to soften the visual impact, aid stability and determine the final type of land use.

In some places the objective may be to restore the area to close to its original condition, with all its environmental and heritage or conservation values intact. In the arid shrub zone this would be rarely achievable and is an unrealistic objective.

SITE DESCRIPTION

Mining development proposals include the information which provides the standards for later rehabilitation, as sites are described in terms of land form, geology, soil types,

Successful rehabilitation of a mine site in the Murchison

surface and ground water behaviour, flora and fauna components and climate, particularly rainfall. This information governs the objectives for revegetation developed in the mine rehabilitation plan.

Mining will result in changes to the physical environment. These changes can be managed so as to avoid adverse environmental impacts, provided that there is information on such matters as surface and deep soil drainage, the susceptibility of the soil surface to erosion and the fragility or otherwise of the local vegetation.

Characterisation of overburden and waste produced by the mine is also important, as some waste rock is hostile to plant growth, being highly alkaline or else giving rise to very acidic breakdown products. Some waste may be very saline. These difficult materials, when identified at an early stage, can be placed selectively in waste dumps. However, placement of difficult materials may be impossible or too costly if identification occurs at a later stage in production.

SITE PLAN

Many of the potentially adverse effects of mining operations can be avoided or reduced by careful siting of infrastructure. The preparation of a site plan enables the clear identification of the important physical elements of the site, including environmentally sensitive locations.

The Use of Vegetation in the Rehabilitation of Mine Sites

The success of the revegetation program will depend upon the attention which is given to:

- the development of the final land form;
- measures which will reduce wind and water erosion on newly planted slopes; and
- the selection, storage and use of topsoil.

THE FINAL LAND FORM

The reshaping and grading of a site is an essential aspect of rehabilitation. Slopes must be stable in order to retain respread topsoil and to sustain subsequent re-establishment of vegetation.

Slopes of 20° or less, with a slope run of 50–60 m between benches, are best for revegetation, due to their capacity to retain water on-site for plant growth. With steeper slopes, water runs off. Moonscaping on steeper slopes is an acceptable alternative where it is feasible and complies with other constraints. Moonscaping holds water in niches for the growth of plants.

THE REDUCTION OF WIND AND WATER EROSION

While a vegetation cover is being established, there are three basic methods of controlling wind erosion (i.e. reducing wind velocity at the soil surface):

- protection of the surface using a mulch;
- leaving the surface rough, as described below for water erosion; and
- establishing windbreaks.

Cover crops of cereal rye were previously recommended to prevent wind erosion. Indigenous annuals such as annual saltbush and pop saltbush are now preferred to exotic species such as cereal rye.

Intense thunderstorms are a feature of mining areas. Water erosion will result unless measures are taken to reduce the effects of such storms. These include:

- The maximisation of infiltration on waste dumps and other treated surfaces by deep ripping to alleviate compaction. Always rip deeply and on the contour, at a spacing approximately equal to the depth of ripping.
- The construction of water-handling structures to carry water off the treated faces when there are heavy storms. Back-sloping benches and vertical drains can be used to accommodate the excess run-off.

TOPSOIL

Topsoil is a valuable resource and must be managed and husbanded. It contains seeds, soil organisms and nutrients, all vital to the successful re-establishment of vegetation.

Topsoil layers are usually about 15–20 cm thick, though most of the nutrients and seeds are in the top 2–3 cm. Below 20 cm the soil lacks fertility and may not, for other reasons, such as low pH, encourage the growth of plants.

In some instances, particularly in the hardpan plains, topsoil is a scarce commodity. If this is the case, reserve the topsoil for the areas that need it most. Do not spread thinly. Ideally, a layer about 15 cm thick should be applied.

The following points should be considered when topsoil is reserved:

- Always save as much topsoil as is possible.
- Direct replacement of recovered topsoil on an area awaiting rehabilitation will give the best results. This prevents or reduces the deterioration of the biological components in the soil which occurs during storage, and also saves double-handling the soil.
- If topsoil must be stockpiled, plan to re-use it as soon as possible. Do not stockpile it in large heaps. The piles should be less than 1–2 m high. Revegetate the stockpile to protect it from erosion, to discourage weeds and to retain soil microbial populations.
- Do not strip topsoil while it is wet, as this can lead to compaction and a loss of structure.
- Augment topsoil with an underlay of subsoil with reasonable properties for plant growth if topsoil is limited.
- Avoid 'borrowing' topsoil.
- After the topsoil is spread, deep rip the surfaces to knit the topsoil into the dump or dam faces.

THE SELECTION OF PLANT SPECIES

In the arid zones of Western Australia, re-establishment of a diverse and permanent cover of local species is generally the rehabilitation aim.

Before species are selected, soil conditions should be assessed in case there are any serious soil deficiencies which need to be rectified with fertiliser. Difficult sites should also be identified at this time and species which have the best chance of survival should be then selected.

Generally seeds of locally growing plants are preferred, even though commercial seeds of a particular species may be available. The former are well adapted to the prevailing climate and soil conditions at the site.

Within a single species, a wide range of ecotypes has evolved in response to site conditions. As a consequence, seeds of a species from one location may not necessarily thrive when used elsewhere. Collection of seed from areas near the mine site is strongly recommended.

When selecting suitable plant species the following points should be considered:

- Observe plant species growing naturally on old disturbed areas near the rehabilitation site so that good colonising and follow-up species can be identified.
- Observe the soil and drainage conditions to which the different local species are adapted and match them with the conditions on the mine site.
- Select plant species that produce sufficient viable seed so that it may be harvested economically.

- Consider habitat requirements of wildlife in the area and choose a mix of species, low growing and tall growing, as well as a mix of plant families, such as saltbushes, poverty and fuchsia bushes and acacias.

In those instances where the area to be vegetated has extremes of acidity or salinity, it will be necessary to look beyond the immediate local environment for suitable species.

SEED ACQUISITION

Seeds can be bought commercially or collected by site personnel. Where the decision has been made to collect local seed, licences to harvest seed must be obtained. Seeds may have to be collected from several areas to match the range of site conditions on the mine site.

Site personnel should understand the correct procedures for seed collection to ensure seed is viable (i.e. ensure it is mature when collected, unaffected by insects and correctly dried before storage).

If buying seed commercially, ensure that seed has been collected locally, stored correctly and is not old. Some seeds lose viability within 12 months.

All seed, whether locally collected or bought from commercial sources, should be tested for germination rate before being committed to a reseeding program. As a general rule, the seeding rates recommended in this book should be doubled if the germination rate is lower than 30 per cent.

SEED STORAGE

Seed should be cleaned, dried, labelled and protected from insect attack. Winnowing in a draught of air allows the separation of much of the vegetable matter from the seed. Particular care needs to be taken to remove leaf material, otherwise fungal growth can reduce seed viability. Prolonged drying in a warm, sunny position, with frequent turning, will be necessary. For most species, storage at less than 10 per cent humidity in the coolest temperature available will maintain the viability of seed.

Insect damage can be controlled by mixing a pest strip (e.g. a chopped pet flea collar) with the seed in storage containers.

The natural life of seed varies with species. Acacia seed is very long-lived — more than five years — while saltbush and bluebush seed lasts for two to three years provided that it is dry and held under ideal conditions.

SEEDBED PREPARATION

Seedbed preparation involves the roughening of the surface to provide niches in which seeds can lodge and germinate. The niches also locally increase water supplies for the establishing plant. Normally, the seedbed will have been created by contour ripping after the topsoil has been spread, as described in the section on topsoil.

There are a number of points which need to be considered when the seedbed is prepared:

- Time the ripping and seedbed preparation so that it takes place just before the onset of the most reliable seasonal rain. In the southern pastoral areas below the Ashburton, rip in March or April to catch the most reliable rains of May–June.

In the more northerly areas, ripping should take place just before Christmas. Ripping in late winter and spring is not recommended.

- Do not work the seedbed or prepare it when the soil is wet and sticky as the working of wet soil creates conditions suitable for erosion.
- Apply the fertiliser, if necessary, during contour furrowing operations. Fertiliser rates should be about 100 kg Agras-Mo per hectare. Higher levels are not recommended as they encourage weed growth which adversely affects plant establishment.

HOW MUCH TO SEED AND WHAT TO SEED

Broadcast 6–10 kg of seed per hectare. Most importantly, this should be a mix of species, including annuals for early ground cover, shrubs of all sizes and some low tree species. A potential mix in the southern areas south of the Pilbara could include:

25%	Acacia species — usually rapidly growing and low stature wattles
25%	Saltbushes — a quarter of this portion should include annual saltbushes gathered around the mine site
25%	Bluebush — as many species as possible
5%	Eucalypts (only if common in the area)
20%	Other species such as sheoaks, native poplar, native willow, cassias, broom bush and fuchsia bush, depending on the site.

In the Pilbara, seed mixes consisting of spinifex, perennial grasses, with cassia and acacia from the area, are preferred. The seed mix used will vary from site to site. Irrespective of location, the mix must include a number of different growth forms which will give stability to the plant community being established.

'Bush litter' gathered by sweeping up nearby surface litter can always be used to advantage as it contains a very wide range of annual and perennial species. 'Bush litter' would be in addition to the formal seed mix applied.

The acacia group is not tolerant of weed competition and should only be included if the area is reasonably free of weeds.

Broadcast seeds should lodge in crevices in the freshly prepared seedbed and should not need covering. If the surface is very smooth, it may be necessary to cover the seeds with a very light harrowing.

PRE-SOWING TREATMENTS

The acacia and cassia seeds will require light scarification before sowing, to increase the germination rate. Not more than 50 per cent should be treated, however, in order that some seeds will remain for germination in future years.

Other species do not require pre-treatment.

SUPPLEMENTARY TECHNIQUES

Sowings can be supplemented by spreading the ripped surfaces with a light mantle of shrub brushwood which still retains seed pods and capsules. With time, the pods will crack open and disperse the seed, which germinates readily with the protection afforded by the brush.

PLANTING SEEDLINGS

The advantages of seedling planting include efficient use of available seed, control over species mix and placement, and no limitation on species included.

The disadvantages include higher costs for planting, the cost of nursery operations or purchase of seedlings, checks in growth rate, and in most arid land cases, the necessity of irrigation.

FENCING

The area treated should always be fenced to exclude grazing by domestic and feral animals. In areas where rabbits are common, it is particularly important that they be controlled.

MONITORING AND MAINTENANCE OF RESTORED MINE SITES

It is essential to monitor the success or otherwise of the rehabilitation program and be prepared to re-work any areas of rehabilitation not performing adequately. Adequate funds should be allowed for this monitoring and maintenance.

In Australia there are no recognised criteria for determining when rehabilitation is complete. Handing back responsibility for the management of rehabilitated land to the relevant land owner or government instrumentality should occur when the area is in a self-sustaining, maintenance-free condition, or when an end-point satisfactory to all parties has been reached.

Maintenance that may be required in addition to rehabilitating any failed areas includes:

- watering planted seedlings;
- fencing to prevent grazing; and
- pest and weed control.

ARID LAND REHABILITATION

CHOICE OF SPECIES

Selecting the correct species for regeneration will depend on several factors including the purpose of regeneration, the soil type and the position in the landscape.

PLANTS FOR GRAZING ANIMALS

If the purpose of regeneration is to provide grazing for domestic stock, then hardy, long-lived, palatable species should be chosen. Saltbush, bluebush, and wattles are well suited to regeneration for grazing.

Heavy textured saline soils (e.g. loamy clays) subject to seasonal flooding can be regenerated for grazing with species such as river or swamp saltbush, silver saltbush and spiny bluebush.

Heavy textured saline soils, which are not seasonally inundated, can be regenerated

Cultivation of severely degraded rangelands in the east Pilbara

Plant regrowth three years later

using the same species and others, such as bladder saltbush and sago bush. River saltbush and spiny bluebush planted on such sites will be drought-prone and less successful unless additional water is harvested by earthworks.

Heavy textured non-saline soils can be regenerated by the same species as those used on heavy textured saline soils, as well as species like cotton bush, felty bluebush and three-winged bluebush. Other desirable forage plants, such as currant bush, have been observed to regenerate naturally on heavy textured soils. Despite this, no success has been achieved by reseeding this species.

Light textured soils (e.g. sands, loamy sands and loams) tend not to be saline and are suited to revegetation for grazing purposes by species such as buffel, birdwood, cotton bush, felty bluebush and three-winged bluebush. Where light textured soils are underlain by heavier textured soils, regeneration with most of the saltbushes and bluebushes is possible.

PLANTS FOR AMENITY AND DUST SUPPRESSION

Regeneration for amenity purposes can be achieved by a wide range of species including those already mentioned. Tree species such as prickly acacia, wait-a-while, curara, mulga, limestone wattle and needlebush are also suited to heavier textured soils.

Species for lighter textured soils would also include those mentioned for the regeneration of grazing land, as well as species such as sandplain wattle, poverty bushes, pebble bush and flannel bush.

SEEDING

TIME OF SEEDING

Where summer rain is most common, areas should be cultivated and seeded before rain is expected. Successful regeneration is often ensured if winter rains follow.

Those areas which do not normally receive summer rain should be sown before the onset of winter rains, usually in April. When regeneration is to rely on natural seed spread from existing plants, cultivation should occur before seed begins to ripen and is shed.

SEEDING RATES

The rate of seeding depends on seed quality, type of seed (size, density, etc.) and price of seed.

Rates will vary depending principally on the germination percentage, purity of seed and the size of seed (number of seeds per kilogram). For example, if seed is more than 80 per cent pure and greater than 80 per cent germinable, then rates of up to 4 kg/ha would apply to buffel and birdwood grass, and up to 2 kg/ha for saltbushes and bluebushes. If the quality of seed is less than this, however, increase the seeding rates according to the reduction in quality. Rates would be doubled if seed were 50 per cent pure, with a germination rate of 40–50 per cent.

It is preferable to sow a mixture of species. As species differ in their requirements for germination and establishment, and as climatic conditions are unpredictable, sowing a mixture will ensure that some will succeed.

When mixtures of species are sown, the total sowing rate for shrubs, for example, is up to 2 kg/ha.

SEED TREATMENTS

The germination of acacia seed is enhanced by seed treatments, such as the scarification of the seed coat with sandpaper or immersion of the seed in hot water (95–100°C) for a few minutes.

Saltbush, bluebush and cotton bush seed requires no treatment before sowing. If theft of seed by ants is likely to be a problem, treat the seed with an insecticide such as Maldison.

STORAGE AND COLLECTION OF SEED

See the section on mine site rehabilitation.

LAND PREPARATION AND SEED PLACEMENT

DEEP FURROWING AND BANKING

Furrowing has been used in other parts of Australia (particularly New South Wales) to regenerate deep, friable soils. Banks have been constructed in Western Australia with a grader blade to create a bank of earth on the downslope side and a furrow on the upslope side.

The grader blade turns soil loosened by ripper tines, in a single-pass operation. The bank provides niches for seed entrapment and forms a barrier to run-off. The furrow collects run-off to enhance subsoil moisture storage.

Furrowing is conducted on the contour bank. If the ends of the banks are turned upslope slightly, ponding behind the bank will occur, facilitating further infiltration.

SEED PLACEMENT

To maximise germination and establishment, seed must be placed in favourable niches. Generally, this means directing seed onto areas of the cultivation with the greatest soil moisture (or potential for greatest soil moisture, when sown dry).

Saltbush and bluebush will not germinate if seed is buried by more than a few millimetres of soil. It is safer to place the seed on the top of cultivated ground. Acacia seed will germinate satisfactorily if sown onto a roughened surface.

If mulches or surface stabilising sprays are necessary to ensure reliable establishment, place them over seed on top of cultivated soil. The cost of seeding with these soil amendments is more expensive than seeding alone, but establishment is often greatly enhanced.

With ripping, favourable sites for germination are adjacent to the rip lines, but not in them. On ridging, seed should be placed on the lower sides of the banks at the upper level of water ponding, and on the uncultivated margin of the downslope furrow.

With pitting, place seed into the pits. As the pits fill with water the lighter, fluffier saltbush and bluebush seeds float and when the water level recedes they become stranded on the upper faces of the pit.

Seed placement for furrowing is similar to that for ridging. Concentrate the seed on the upslope side of the bank, slightly above the old ground level to where water in the furrow will pond.

For all methods of cultivation, avoid placing seed where it will become waterlogged, causing seedlings to die.

Section I

Annual herbs,
short-lived perennials,
ferns and vines

Toothed saltbush

Atriplex acutibractea

Family: Chenopodiaceae

GENERAL DESCRIPTION: Toothed saltbush is a low, grey, short-lived, perennial shrub which grows to 50 cm, with stiff, upright, tangled branches. The leaves are oval, 1–2 cm long and are edged with broad, shallow teeth. The margins curl up while the tips curl down. Very small male and female flowers are borne on the same plant, unlike other perennial saltbushes which have separate male and female plants. The female flowers develop into narrow, three-toothed seed cases about 3 mm in diameter and 4–8 mm long.

HABITAT: Toothed saltbush is a species of disturbed areas and grows on a variety of soil types. It is particularly common on roadsides.

FORAGE VALUE: It is not known to be grazed by stock.

Toothed saltbush, about 40 cm high

INDICATOR VALUE: Its presence in large amounts, year after year, indicates a heavily overgrazed rangeland.

REHABILITATION: Toothed saltbush could be included in a mix of annual species used to colonise waste dumps. The seed would normally be obtained when 'bush litter' is gathered. It is very tolerant of alkaline and saline conditions. Its lack of attractiveness to stock makes it a suitable species for early colonisation of waste dumps.

Fruit, about actual size

Dwarf saltbush   Bell saltbush, Flat-topped saltbush

Atriplex codonocarpa

Family: Chenopodiaceae

GENERAL DESCRIPTION: Dwarf saltbush is an annual, or in rare cases a short-lived perennial, which grows to 30 cm, but occasionally reaches 60 cm. It is rounded, intricately branched and is a mealy, bluish grey. The leaves are angularly toothed and up to 3 cm long. The seed is carried in a spongy structure resembling a broad-based cone or bell; the flattened part being uppermost and the apex forming the attachment to the plant. The seed cases are produced very freely and plants often appear to be completely covered by them. Dwarf saltbush has male and female flowers on the same plant.

HABITAT: Dwarf saltbush is found in saltbush pastures. It is a successful coloniser of disturbed areas, such as roadsides, which receive extra run-on water. It occurs on loams and clays, but avoids sandy soils.

Dwarf saltbush, about 15 cm high

FORAGE VALUE: Dwarf saltbush is usually unpalatable, though stock have been observed to eat it when pressed.

INDICATOR VALUE: It indicates poor range condition when it dominates saltbush pastures of the river plains, lake country and saline, stony uplands.

REHABILITATION: Seed of dwarf saltbush can be gathered easily in most degraded saltbush communities and should be included in the 'bush litter' component of the seed mixture applied to mining dumps. It is tolerant of saline and alkaline conditions and is often the first natural coloniser of waste dump slopes.

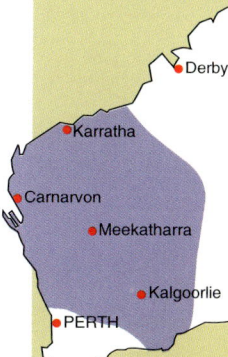

Almost mature fruit, about actual size

Pop saltbush

Atriplex holocarpa

Family: Chenopodiaceae

GENERAL DESCRIPTION: Pop saltbush is a rounded, grey, mealy, annual herb, which rarely exceeds 30 cm and is more commonly 15 cm high. The leaves have an irregularly notched margin, are about 3 cm long and turn pinkish towards maturity. Separate male and female flowers are carried on the same plant. After the seed is set the seed case develops into a lemon-shaped, or globular, spongy mass about the seed. At maturity the seed structures turn blackish and, falling off, form a crisp, dense mat on the ground. It is distinguished from dwarf saltbush, which also has a spongy fruiting body, by its globular or oval-shaped fruit.

HABITAT: Pop saltbush is found in more salty situations than is dwarf saltbush. It generally grows on the lower parts of salt lake margins, saline drainages and

Pop saltbush, about 20 cm high

on saline, rocky hills where the soil tends to be heavier in texture. It favours areas receiving run-on water and is never found on sands.

FORAGE VALUE: It is not a preferred species and stock eat it only when forced to do so.

INDICATOR VALUE: Pop saltbush is an indicator of poor range condition when it dominates the communities in which it grows.

REHABILITATION: Pop saltbush is very tolerant of saline conditions. Mixtures of 'bush litter' containing it may be sown on the tops of saline tailings dams and on the more saline waste dump slopes, provided that it is possible to create some drainage to reduce salt locally. It is unpalatable and a useful coloniser of very difficult situations. It will normally be gathered separately from dwarf saltbush as it grows in more saline situations.

Almost mature fruit, about actual size

Annual saltbush

Atriplex semilunaris

Family: Chenopodiaceae

GENERAL DESCRIPTION: Annual saltbush is a spreading, robust annual which can reach a height of 50 cm and is often 1 m wide. The leaves are a bluish green and have a scaly sheen. They are about 4 cm long and 1 cm wide and have an irregularly notched margin. The seed cases are borne freely, are flattened and more or less semi-circular and have a coarsely toothed edge. They are found along the stem and in terminal spikes. They drop freely at maturity.

HABITAT: Annual saltbush grows in saltbush country, on saline valley floors and in disturbed areas which are slightly saline or alkaline. It favours loams and clays.

Annual saltbush, about 20 cm high and 80 cm across

FORAGE VALUE: It is unpalatable and is rarely eaten by stock except when they are forced to do so by lack of other feed. Being an annual, it does not persist and cannot be regarded as a useful fodder plant.

INDICATOR VALUE: When it dominates a pasture, annual saltbush is an indicator of poor range condition. It is a successful coloniser of disturbed sites where it is frequently found in abundance.

REHABILITATION: Annual saltbush is less tolerant of saline conditions than dwarf and pop saltbushes are and should only be sown in the less saline waste areas. It should colonise slopes covered with deep topsoil. The seed is collected along with the whole plant but as the seeds are shed easily it may be necessary to sweep or vacuum them up from the surface.

Fruit, about actual size

Parakeelya Four-o'clock

Calandrinia polyandra

Family: Portulacaceae

GENERAL DESCRIPTION: Parakeelya is a succulent, prostrate annual with a large and deep tap root. The leaves are thick and fleshy, narrow, a light green and vary in size from 2.5 cm to more than 7.5 cm long. As they mature they turn a dull reddish brown. The flowers are very showy, 2.5–4 cm across and are borne in loose groups at the ends of erect stems up to 15 cm long. Each has five pinkish, shining petals, a yellow centre and numerous stamens. They open for one day only. The seeds are very small and numerous and are contained in a pointed, thin-walled capsule.

HABITAT: Parakeelya is found in a great variety of habitats in floodplains, in shallow loams beneath granites and on broad stony plains. It grows with mulga, bowgada, Wilcox bush and tall sida.

Parakeelya, about 20 cm high

FORAGE VALUE: It is readily eaten by stock. It provides forage even after the soil moisture has dried up since it can live, flower and seed on the moisture held in the leaves. It contains high levels of an oxalate-forming material, which leads to the formation of urinary calculi in wethers and rams. When these sheep graze pastures consisting largely of parakeelya, urinary stones form and block the narrow urethra of the male causing urine retention, which ultimately ruptures the bladder. There is no known cure. However, this is not a common event. When parakeelya is abundant, male sheep should be transferred to paddocks where there are other ground feeds in addition to parakeelya.

INDICATOR VALUE: It has no known indicator value.

Flowers and seed pods, about actual size

Ward's weed

Carrichtera annua

Family: Brassicaceae

GENERAL DESCRIPTION: Ward's weed is a multi-branched, annual herb, which grows to 40 cm. The leaves are branched into many apparent 'leaflets', the whole being up to 5 cm long. The plant is characteristically bristly. The flowers have four pale yellow petals with striking purple veins. Flowers are carried in long racemes and are succeeded by short, roundish pods which have long, compressed beaks. The seeds are dark brown in colour and about 1 mm in diameter. The plant has a definite pungent odour, particularly when crushed and after rain has fallen on it. It has been introduced from the Mediterranean region.

Ward's weed, about 40 cm high

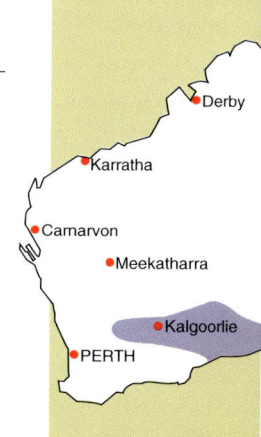

HABITAT: Ward's weed is associated with saltbush and bluebush communities on strongly alkaline soils.

FORAGE VALUE: Only the young plant is grazed by stock.

INDICATOR VALUE: Ward's weed is an invading plant which rapidly colonises disturbed areas and communities weakened by fire and overgrazing. It is, when very prevalent over extensive areas, a reliable indicator of declining range condition.

Perennial tar vine

Commicarpus australis

Family: Nyctaginaceae

GENERAL DESCRIPTION: Perennial tar vine is a straggly, scrambling perennial that can climb up to 3 m and be as much as 4 m wide. The leaves are dark green, opposite each other and heart-shaped. They tend to be fleshy. They are 2–5 cm long, up to 3 cm wide and are borne on short stalks. The flowers are purple, five-petalled, about 10 mm in diameter and with the stamens and style extended from the throat. Three to eight flowers are borne together on a common stalk. The fruit is small and club-shaped, up to 8 mm long and has conspicuous knobby glands which produce a sticky secretion.

Perennial tar vine, about 4 m across

HABITAT: Perennial tar vine is usually found beneath more upright perennials which shelter it from grazing. It grows on clays, loams and on low sandy banks developed on floodplains and on soils produced from sedimentary rocks.

FORAGE VALUE: It is readily grazed after the annual herbs and grasses wither and is a valuable perennial in the pasture.

INDICATOR VALUE: The indicator value is not known. It is a useful guide to the level of use of the current season's growth.

Flowers and seeds, about actual size

Round-leaf pigface

Disphyma crassifolium subsp. *clavellatum*

Family: Aizoaceae

GENERAL DESCRIPTION: Round-leaf pigface is a prostrate perennial creeper that grows to 10 cm in height and 50 cm in diameter. The prostrate stems root at the nodes and hold their leaves in clusters. Each leaf is 4–6 cm long, 4–6 mm thick and succulent. The pink or purple flowers are about 4 cm in diameter and consist of numerous 1.5 cm long petals attached to a woody base. The base develops into a seed capsule with five flaps that open up to release small seeds.

HABITAT: Round-leaf pigface grows on semi-saline clays marginal to salt lakes in association with bladder saltbush, five-pin bluebush, samphire and other salt tolerant shrubs.

Round-leaf pigface, about 50 cm across

FORAGE VALUE: It is rarely eaten by stock as it contains high quantities of salt.

INDICATOR VALUE: It is an intermediate species that has been known to increase in some, but not all, overgrazed situations.

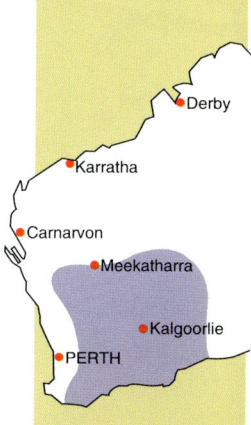

Flowers, about half actual size

Cannon balls   Curious bassia, Ball burr

Dissocarpus paradoxus

Family: Chenopodiaceae

GENERAL DESCRIPTION: Cannon balls is a low, semi-perennial plant which grows to 30 cm. It is covered in white, woolly hairs and has large, spiny, woody seed capsules. The leaves are 5–15 mm long, and are straight and succulent. Eight to ten insignificant flowers are produced in the junctions of the leaves and stem. These later unite to produce the woolly fruiting body or 'cannon ball' which is actually a compound fruit about 1 cm in diameter. After flowering, the ball becomes woody, loses its woolly hairs and exposes many short stiff spines. The fruits often litter the ground for many years after being shed.

HABITAT: Cannon balls grows on limestone, slightly saline floodplains and around salt lakes. It is found with river saltbush, limestone poverty bush, limestone wattle and tall saltbush.

Cannon balls, about 20 cm high

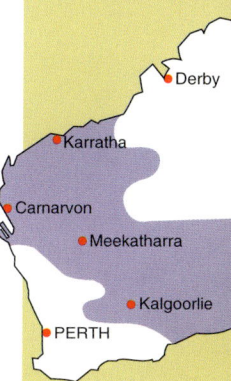

FORAGE VALUE: When green, it has a protein content of about 13 per cent and is readily eaten, but it is not eaten when dry and woody. It has the ability to sprout rapidly after rain and for this reason can contribute significantly to the forage on offer.

INDICATOR VALUE: It increases in overgrazed shrublands where it takes the place of perennials such as river saltbush and sago bush. When found in abundance to the exclusion of more robust perennial plants, it indicates poor range condition.

REHABILITATION: Cannon balls could be used as a colonising plant on waste dump and tailings dam slopes, though seeds will rarely be found in large quantities. It has the significant advantage of weak perenniality over the annual saltbushes and would, therefore, stabilise the slopes more effectively.

Mature and immature fruit, about actual size

Green crumbweed

Dysphania kalpari

Family: Chenopodiaceae

GENERAL DESCRIPTION: Green crumbweed is a bright green, low-growing, annual herb which bears vertical seed heads above the somewhat sticky, prostrate, vegetative growth. It can reach 30 cm and be 40 cm wide, but commonly is less. The green leaves are ovate and up to 2.5 cm long, with lobed and wavy margins. When crushed they have a strong aromatic odour. The flowers are held in long 'rat tail'-like spikes above the leaves. Each fruiting body is about 1–2 mm in diameter with four greenish lobes and contains a shiny, blackish seed. The seed heads pale from green to yellow on maturity.

Green crumbweed is very conspicuous because of its bright colour, but it never forms a total ground cover in the communities in which it occurs. It grows as single plants or in small clumps.

HABITAT: Green crumbweed grows on shallow soils with hardpan, on shallow sandy banks and on sandplains. It avoids heavy clay soils.

Green crumbweed, 40 cm across

FORAGE VALUE: It is not grazed by stock to any extent and it is reported to react positively to tests for cyanide. Closely related species have been implicated in outbreaks of stock poisoning following summer rains in the eastern wheatbelt of Western Australia.

INDICATOR VALUE: It has no known indicator value in the assessment of range condition.

NOTE: A closely related species, *D. rhadinostachya*, illustrated below, has a many-branched, pyramid-like flower head.

Flowers, about actual size, of the closely related species D. rhadinostachya

Woolly bindii   Woolly-fruited copperburr

Eriochiton sclerolaenoides

Family: Chenopodiaceae

GENERAL DESCRIPTION: Woolly bindii is an erect, short-lived perennial which grows up to 25 cm. The leaves are narrow and placed alternately along the stem. They are 5–10 mm long, round in cross-section and covered in long, white hairs that are pressed flat onto the leaves. The flowers are very small, and develop into a seed capsule encased in dense, woolly, white hairs from which extend five soft wings in a star formation. The star-shaped fruits are about 1 cm across, and from the centre, five erect appendages arise, which are also covered in woolly hairs.

HABITAT: Woolly bindii grows on plains with alkaline, loamy soils variously with salmon gum, pearl bluebush, cotton bush and mulga.

Woolly bindii, about 20 cm high

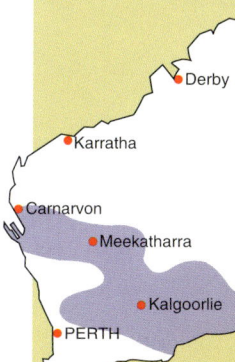

FORAGE VALUE: This species is readily eaten by stock when green and when the annual forage has dried off. Pasture managers should aim at encouraging its persistence, since it produces forage on small amounts of rainfall which are insufficient to germinate and establish annual species. It contains up to 18 per cent crude protein and 10 per cent salt.

INDICATOR VALUE: It is a short-lived species which can be eliminated by severe droughts. It has no value in the assessment of range condition. However, it is a useful guide to the level of use of the current season's production.

REHABILITATION: Woolly bindii is a suitable species for mine dump reclamation. It is perennial and, growing close to the ground, it protects the surfaces from wind and water erosion. It is collected in association with other ground species. It is tolerant of very alkaline conditions.

Fruit with soft spines covered in woolly hairs, about actual size

Crowsfoot   Corkscrew

Erodium crinitum

Family: Geraniaceae

GENERAL DESCRIPTION: Crowsfoot is a hairy annual growing to 50 cm, but more commonly much less. It rises from a fleshy tap root and spreads over the ground covering 1 m or more in diameter. The green leaves are very deeply lobed into three toothed portions. They are carried on slender stalks. The flowers have five, deep blue petals about 1 cm long and up to 5 mm across. They occur in groups of up to six on stalks, which arise in the axils of the leaves and stem. The fruiting body has a tapering beak up to 7 mm long, which splits at maturity into five sharply pointed seed cases. Each seed case is surmounted by a long, spirally coiled awn which is covered with long, white hairs.

Crowsfoot, about 40 cm across

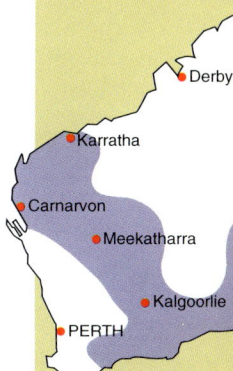

HABITAT: Crowsfoot grows in most soil types, but is more prevalent in mulga communities and on sandy loams and loamy sands. It tends to avoid saltbush communities and clays.

FORAGE VALUE: It is abundant following winter rainfall, often crowding out other species with its rapid growth. It is relished by stock even when dry and is considered to be an excellent feed. The awns of the seeds can cause considerable damage to the skin of sheep and may create sites for subsequent blowfly attack.

INDICATOR VALUE: It has no value as an indicator of range condition.

Seed case, about actual size

Balsam   Mat spurge, Caustic weed

Euphorbia drummondii

Family: Euphorbiaceae

GENERAL DESCRIPTION: Balsam is a prostrate herb which clings tightly to the ground. Individual plants may be up to 30 cm across. The stems exude a milky sap when cut or broken and are reddish in colour. The leaves are opposite each other and are small, about 3–5 mm in size and sub-rounded in shape. They are dark green with a purplish tint which is emphasised with the onset of moisture stress. The margins of the leaves are reddish. The flowers are very small and scattered throughout the plant.

HABITAT: Balsam grows in a wide range of habitats; on river levees, alluvial plains and on red sandplains.

FORAGE VALUE: It is quite palatable and is commonly eaten without ill effects when mixed with other species. Travelling sheep or sheep which

Balsam, about 30 cm across

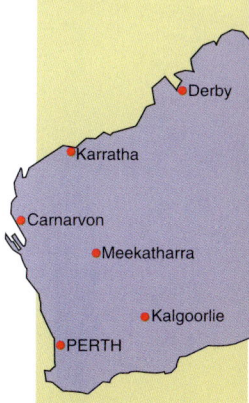

have been held for some time without food have been poisoned by balsam in eastern Australia. There is no record of major stock mortality in Western Australia, though care should always be exercised when the plant is prevalent.

INDICATOR VALUE: It has no indicator value.

Seeds and flowers, about actual size

Poison morning glory

Ipomoea muelleri

Family: Convolvulaceae

GENERAL DESCRIPTION: Poison morning glory is a perennial trailing vine with stems up to 1.5 m long. The leaves, which are scattered along the stems, are heart-shaped and about 2.5–5 cm long. They may be either hairy or hairless. Pink, trumpet-shaped flowers are produced alone or in small groups along the stem. They are about 5 cm in diameter and are succeeded by globular-shaped seed cases 1–2 cm long containing four large, densely haired seeds.

HABITAT: Poison morning glory grows on floodplains and other areas that receive run-on water on to clays and loamy soils. It grows with river red gum in open grasslands or in very sparse shrublands.

Poison morning glory, about 2.5 m across

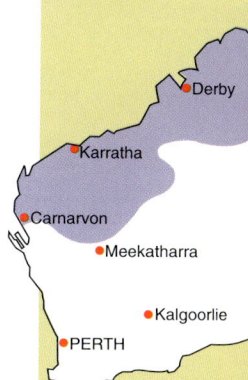

FORAGE VALUE: It grows rapidly after floods and may be eaten then when little else is available.

This species has been responsible for heavy sheep losses in Western Australia. The symptoms described are a steady deterioration in condition with apparent loss of the use of the hind limbs. When driven, many sheep develop a jerky gait and tire easily. They show difficulty in breathing and a severe 'tongue-ing'. Weaners are more seriously affected than adult sheep and die after marked salivation. The toxic principle is in the same group as the hallucinogenic drug LSD. No effective treatment is known for the poisoning. Removal of stock during periods when this species is the major source of forage is the only management practice which can be adopted.

INDICATOR VALUE: It has no known indicator value.

Flower is 5 cm wide

Cottony bluebush   Rope plant

Maireana carnosa

Family: Chenopodiaceae

GENERAL DESCRIPTION: Cottony bluebush is a perennial herb with a woody rootstock. It grows to 50 cm, but is more commonly 15–20 cm high. It has straggly, herbaceous stems and greyish green foliage covered with woolly hairs. The leaves are narrow, up to 12 mm long, flattened and semi-succulent. The flowers are borne in winter and spring and are hidden between the stem and leaves. They produce a leathery seed case which is surrounded by a flat, greenish white wing about 8 mm in diameter, covered with long, white, silky hairs. The seed cases are arranged along the whole length of the stems and give each stem a rope-like or twisted appearance.

Cottony bluebush, about 15 cm high

HABITAT: Cottony bluebush is found on sandy banks, plains marginal to streamlines, on the levees and levee back plains and numerous other habitats. While it occupies slightly saline habitats, it is never found in highly saline country.

FORAGE VALUE: It is palatable to stock, particularly in the young growth stage. It has little drought tolerance, though it will outlast other annual herbage species which commonly grow with it.

INDICATOR VALUE: Cottony bluebush has no known indicator value.

REHABILITATION: Cottony bluebush is a valuable colonising plant on the slopes of tailings dams and waste dumps where these are not highly saline. Seed can be gathered in the field using a vacuum harvester and is normally spread with other annual species in the 'bush litter' component used in mining reclamation.

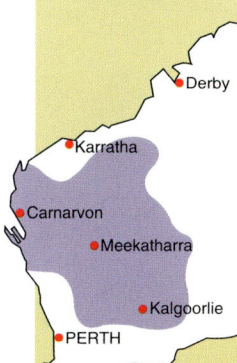

A rope-like spike of almost mature cottony fruits, about one and a half times actual size

Cogla   Native pear

Marsdenia australis

Family: Asclepiadaceae

GENERAL DESCRIPTION: Cogla is a slender climbing plant which exudes a milky sap when damaged. The leaves are opposite, 4–10 cm long, thin and leathery and a mid-green in colour. The flowers are a creamy white and bell-shaped. They are borne in dense clusters at the ends of the stems. The fruit consists of a thick-walled, green, tapering pod, which is 4–10 cm long and covered in dense hairs. On maturity the pods split and release black seeds, each of which has a large tuft of silky hairs.

Mature vine

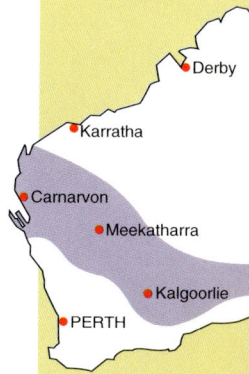

HABITAT: It is sparsely scattered throughout the mulga woodland. It is never common in any location.

FORAGE VALUE: The forage value is not known.

INDICATOR VALUE: It has no indicator value.

NOTE: The fruits are edible when immature. The outer casing becomes hard, fibrous and unacceptable as it ages.

Fruits, about half actual size

Flowers, about actual size

Nardoo

Marsilea drummondii

Family: Marsileaceae

GENERAL DESCRIPTION: Nardoo is a perennial fern with leaves closely resembling those of clover in shape. There are four leaflets borne at the ends of the stems. They are up to 2 cm in diameter and rounded at the tips. A 'nutlet' forms at the base of the leaf stalks. It is 4–9 mm in diameter and covered in dense brown hair.

HABITAT: Nardoo grows in areas subject to inundation such as crabholes, permanent pools and claypans.

Nardoo, about 20 cm across

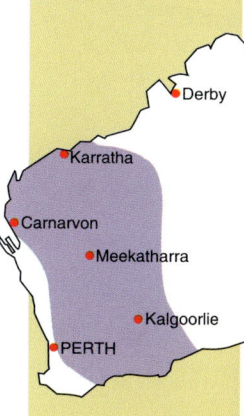

FORAGE VALUE: The foliage is not known to be eaten by stock.

INDICATOR VALUE: It has no indicator value.

NOTE: Aborigines ground the nardoo 'nutlets' to make flour.

'Nutlets' and developing fronds, about twice actual size

Mat mulla mulla

Ptilotus aervoides

Family: Amaranthaceae

GENERAL DESCRIPTION: Mat mulla mulla is a prostrate annual with reddish stems, dark green foliage and brownish flower heads. The plant has a small rootstock and the growing tips and leaf axils are covered in white hairs. The leaves are oval-shaped and 3–25 mm wide. The flowers are carried in dense spikes up to 3 cm long and 1 cm wide. The spikes change from creamy green to pink and brown as they mature.

HABITAT: Mat mulla mulla is common in open plains, underlain by shallow hardpan. It grows with horse mulla mulla and stunted mulga.

Mat mulla mulla, about 60 cm across

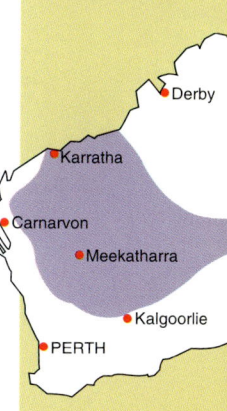

FORAGE VALUE: This plant is highly sought after by both sheep and cattle.

INDICATOR VALUE: It has no indicator value.

NOTE: There are a number of other prostrate mulla mullas, some of which are perennials.

Flowering spikes, two-thirds actual size

Purple mulla mulla   Tall mulla mulla

Ptilotus exaltatus

Family: Amaranthaceae

GENERAL DESCRIPTION: Purple mulla mulla is an erect annual species which grows to 1 m. Upon germination, it initially forms a rosette and later produces stout and erect, branched flowering stems which are up to 1 cm thick. The leaves are bright green, spear-shaped, shiny, fleshy and up to 20 cm long and 7 cm wide. The purple flower heads are initially conical, but become cylindrical and pale mauve on maturity. In good seasons they present a colourful show when massed together.

HABITAT: Purple mulla mulla is found in a wide variety of habitats: on shallow soils on the foothills of sandstone or granite hills, on soils derived from limestone, on alluvial plains and on the less saline sections of river floodplains and lake systems.

Purple mulla mulla, about 80 cm high

FORAGE VALUE: It is palatable and remains green much longer than other annual species. The green plant may contain 16 per cent crude protein, 23 per cent crude fibre and 8 per cent salt. It is a highly preferred forage of sheep and kangaroos.

INDICATOR VALUE: It has no indicator value.

REHABILITATION: Purple mulla mulla has been seen to volunteer on the slopes of waste dumps and may be presumed to be a useful annual component of the colonising species. There is insufficient information on its germination requirements so it cannot be recommended as a major constituent of the annual mix used in the rehabilitation of sites. It would be worthwhile to have it included, however, as a minor ingredient.

Flower heads, about three-quarters actual size

Pussytail mulla mulla   Featherheads

Ptilotus macrocephalus

Family: Amaranthaceae

GENERAL DESCRIPTION: Pussytail mulla mulla is an erect annual growing to 1 m. The leaves are crinkle-edged and are up to 10 cm long and 4 cm wide. The flower heads are conical when immature, the flowers at the base opening first. They are up to 10 cm long and 3 cm wide. When all have opened, the head has a cylindrical shape.

HABITAT: Pussytail mulla mulla generally grows on loamy soils that are periodically flooded and it is common in good seasons. It rarely occurs on saline country.

Pussytail mulla mulla, about 70 cm high

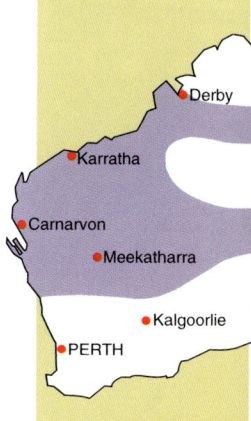

FORAGE VALUE: When green it is palatable and contains up to 11 per cent crude protein, but when dry it becomes fibrous and its crude protein content declines to about 5 per cent.

INDICATOR VALUE: It has no indicator value for range condition and its presence reflects good seasonal conditions.

A flower head, about actual size

Green mulla mulla   Bottle washers

Ptilotus polystachyus

Family: Amaranthaceae

GENERAL DESCRIPTION: Green mulla mulla is a tall, but straggling, annual reaching 1 m, with narrow, green flower heads and rough foliage. The leaves are pale green and up to 18 cm long and 2 cm wide. They are harsh to the touch and the leaf margins are often crinkled. The flower heads are narrow cylinders 10 cm long and 2 cm wide and are carried at the tip of each stem.

HABITAT: Green mulla mulla grows mainly on sandy soils in wanderrie country and on other low sand banks and dunes. It grows beneath mulga

Green mulla mulla, about 90 cm high

and Wilcox bush. It is widespread in good rainfall years and dominates the ground flora, suppressing, by its vigour and height, other species which grow with it.

FORAGE VALUE: Crude protein content may be as high as 22 per cent. It is eaten when green, but when dry it is rough and unpalatable. It is not as durable as purple mulla mulla and wilts rapidly with moisture stress.

INDICATOR VALUE: Green mulla mulla has no indicator value. Its presence in quantity reflects good seasonal conditions.

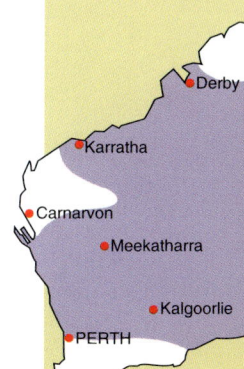

A flower head, about half actual size

Roly poly   Buck bush

Salsola kali

Family: Chenopodiaceae

GENERAL DESCRIPTION: Roly poly is an annual or biennial herb which is normally about 60 cm high and the same across, but it may exceed 2 m in diameter. The young, bright green leaves are round in cross-section, pliable and up to 5 cm long. As the plant matures the leaves become stiffer, somewhat flattened and taper to a sharp point. The flowers are produced between the leaf and stem towards the ends of the branches. After seed set, a seed case develops which has five horizontal papery wings up to 7 mm wide. These resemble the seed cases of many bluebushes. The whole plant often breaks away from the ground intact. It can roll away unimpeded in degraded areas and masses against fences.

Roly poly, about 40 cm high

HABITAT: Roly poly grows in most environments and is particularly common on saline and alkaline soils and limestone banks.

FORAGE VALUE: It is eaten in the early stages of growth, but as it matures it becomes unacceptable to stock.

INDICATOR VALUE: Roly poly is an indicator of poor range condition when it dominates a pasture.

REHABILITATION: Mature roly poly is very brittle and it will in consequence be difficult to collect seed without dried stalks. The mature plants should be gathered whole and spread across dump faces, always ensuring that seeds have not become detached during collection. The plant is recommended for very difficult situations of salinity, alkalinity and aridity. It is unlikely to dominate once more durable species become established.

Fruits and spiny leaves, about actual size

Goathead burr

Sclerolaena bicornis

Family: Chenopodiaceae

GENERAL DESCRIPTION: Goathead burr is a grey, short-lived shrub which grows to 60 cm. The leaves are 10–25 mm long, thin and succulent and densely covered with white, woolly hairs. The flowers are insignificant, but they form fruits with large, 7–8 mm round, woody cases that carry two outward-pointing, aggressive, robust spines, 10–15 mm long.

HABITAT: It occurs in Roebourne Plains grasslands in huge populations following heavy rains. It is also found in saline, clayey areas with prickly acacia.

Goathead burr, about 90 cm across

FORAGE VALUE: When green, it is readily eaten by stock, particularly sheep, but it is not eaten when dry.

INDICATOR VALUE: When seen in large populations it indicates country in fair to poor range condition.

REHABILITATION: Goathead burr is not recommended for rehabilitation purposes as the large, woody seed cases carry spines which can inflict damage to stock. If introduced deliberately, goathead burr has the capacity to spread beyond the mine site onto degraded areas nearby.

NOTE: It appears to establish during good seasons if the stocking pressure is not high and then lives for up to two years. It usually dies during a prolonged dry period.

Seed cases, with woolly outer covering, about two-thirds actual size

Yellow bindii

Sclerolaena cuneata

Family: Chenopodiaceae

GENERAL DESCRIPTION: Yellow bindii is a prostrate and dense annual which grows to 30 cm. The leaves are succulent and green and up to 5 mm long. They are round in cross-section and are generally hairless. The flowers are very small and inconspicuous and are found in the axils between leaf and stem. The seed case has four spines, three long and one short, spreading out from the seed case. The plant seeds freely and this gives it a dense covering of characteristic yellow and aggressive prickles.

HABITAT: Yellow bindii is most common on heavy, mineral-rich soils and on texture contrast soils. It grows with perennial shrubs such as sago bush, Gascoyne bluebush and golden bluebush. It is very common on degraded, sluggish drainages in hilly country, where it replaces more valuable perennial shrubs.

Yellow bindii, about 30 cm high

FORAGE VALUE: It is thought to be of some value before the spines develop on the seed cases. However, in a study on the Gascoyne catchment, yellow bindii was not eaten even though it was freely available. After the seeds mature, the plant rapidly becomes senescent, dropping what remain of its leaves. The dried-out stems and seeds are inedible.

INDICATOR VALUE: It is an increaser species. When it is found abundantly in communities which should support sago bush, Gascoyne bluebush and golden bluebush, it indicates poor range condition.

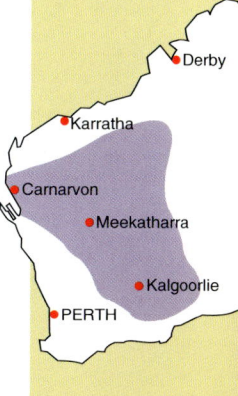

Spiny yellow fruit, about one and a half times actual size

Hairy bindii

Sclerolaena densiflora

Family: Chenopodiaceae

GENERAL DESCRIPTION: Hairy bindii is a short-lived, globular, dark green shrub up to 30 cm high. The whole plant is covered with silky hairs. The straight leaves are 10–20 mm long, fleshy and 2–3 mm wide. Small flowers form in the leaf axils and develop into hard seed cases about 2 mm in diameter. Five relatively inoffensive spines, each 3–5 mm long, are located at the top. The entire seed case is also covered in dense silky hairs.

HABITAT: Hairy bindii is present in most habitats with hardpan soils. It is also associated with gibber-covered plains.

Hairy bindii, about 45 cm across

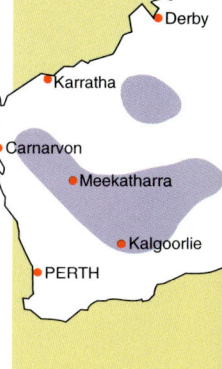

FORAGE VALUE: It is not readily eaten by stock. As it is not persistent during dry periods, it does not provide any drought reserves for stock. At the end of a winter season it dries and is rapidly dispersed.

INDICATOR VALUE: In large populations it indicates poor range condition as it increases in situations where competing shrubs have been overgrazed.

REHABILITATION: It is a useful coloniser on arid sites and would be useful if spread with 'bush litter'. It is unlikely to be successful on strongly alkaline or saline sites and should be reserved for those areas which are neutral in soil reaction and non-saline.

Densely hairy stem, leaves and developing seed cases, about one and a half times actual size

Grey bindii   Grey copperburr

Sclerolaena diacantha

Family: Chenopodiaceae

GENERAL DESCRIPTION: Grey bindii is a prostrate, perennial herb which grows to 30 cm and up to 40 cm in diameter. The branches have a dense covering of grey hairs. The leaves are 10–20 mm long, 3 mm wide, oblong and pointed at the tip. The flowers are very small and insignificant and occur in the axil between leaf and stem. The seeds are held in generally woolly or hairy bodies up to 4 mm across. There are two short, stout spines about 2 mm long on each side of the fruiting body.

This widespread species can vary somewhat throughout its range; the fruiting bodies may, for instance, take on a yellowish colour. Although it is a perennial, it is not long-lived as few plants survive for longer than five years.

HABITAT: Grey bindii is found on a wide variety of soil types and grows with a wide range of species from bluebushes to grasses.

Grey bindii, about 20 cm high

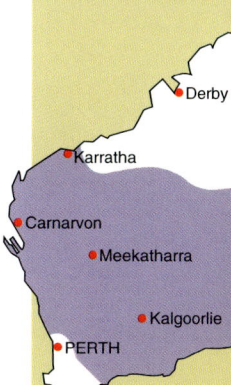

FORAGE VALUE: It is a valuable forage species as it provides good quality feed during dry periods after the annuals have withered. It contains up to 13 per cent crude protein and 10 per cent salt. It is one of the most useful of the bindii group.

INDICATOR VALUE: It is not usually a good indicator of range condition as it is subject to prevailing weather conditions. However, when it is abundant or if it is seen to be producing numerous seedlings, it may be an indicator of good range condition.

REHABILITATION: Grey bindii should be included where possible in seed mixes. It is able to give stability to the stand of plants being established. Seed could be gathered along with the plant itself and spread as soon as possible after collection.

Almost mature fruit with short spines, about twice actual size

Silky bindii   Tall bindii

Sclerolaena eriacantha

Family: Chenopodiaceae

GENERAL DESCRIPTION: Silky bindii is an erect shrub that grows to 60 cm. It has prominent, woolly hairs on the stem and fruit. The leaves are silky and up to 2 cm long. They are succulent and flattish on the upper surface and curved upwards. The flowers are small and inconspicuous. The seed case is about 3 mm in diameter and bears two fine, long spines about 7 mm long. The fine spines just protrude from the covering of hairs on the seed case.

HABITAT: Silky bindii occurs on slopes below breakaways, on flow lines, on plains in basalt country, and on river floodplains. It favours duplex soils, and clays and loams which are generally neutral to alkaline. It grows with golden bluebush, sago bush and frankenia.

Silky bindii, about 30 cm high

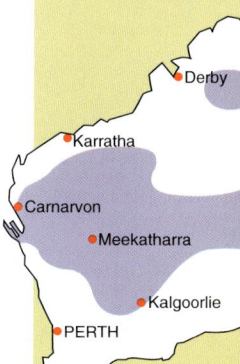

FORAGE VALUE: It is readily eaten by stock. The crude protein can be as high as 20 per cent and the digestibility up to 63 per cent.

INDICATOR VALUE: It is a decreaser species and when prominent in the pasture it indicates fair to good range condition.

REHABILITATION: Silky bindii will be found growing best near breakaways. If seeding has been prolific it could be included in the mix used for colonising dump and tailings dam slopes. It is able to tolerate moderately saline conditions and dry environments. If site conditions indicate the need for a plant with these characteristics, silky bindii should be gathered when the opportunity presents itself.

NOTE: Silky bindii can be distinguished from fluffy bindii by the spines on the fruit. There are two fine spines on silky bindii, but there are no spines on fluffy bindii, only soft filaments.

Woolly, twin-spined fruit, about actual size

Fluffy bindii

Sclerolaena eurotioides

Family: Chenopodiaceae

GENERAL DESCRIPTION: Fluffy bindii is a straggly perennial with weak branches covered in long, white, silky hairs. It reaches up to 30 cm. The leaves are 6–12 mm long, succulent, curve upwards and are round in cross-section. The flowers are small and insignificant. The seed case is 2 mm in diameter and has three to five 10 mm long filaments rather than spines protruding from the mass of white silky hairs. Fluffy bindii is distinctive since it does not possess the sharp-spined fruits common in the bindii group.

HABITAT: Fluffy bindii grows on river floodplains, on slopes below breakaways, on the fringes of salt lakes and on greenstone and basaltic slopes. It occurs on duplex soils and on clays and loams that receive extra

Fluffy bindii, about 20 cm high

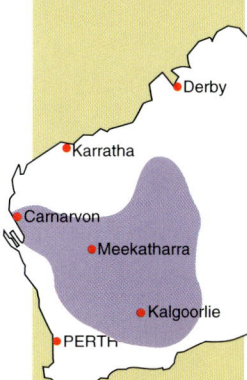

water or have good water-holding capacity. The soils are generally neutral to alkaline. It is frequently found growing inside the protective umbrella of bluebushes or other woody species, which appear to protect it from grazing.

FORAGE VALUE: It is very readily eaten by stock.

INDICATOR VALUE: It is a decreaser species and mainly found in bluebush/saltbush pastures which are in good range condition.

REHABILITATION: Fluffy bindii does not appear to be as robust as silky bindii and is not favoured as much for rehabilitation work. It is found only in locations where there is run-on water. However, if site conditions are favourable it could be included in the colonising mix sown after the slopes have been prepared.

Hairy fruit with soft filaments, about twice actual size

Limestone bindii

Sclerolaena obliquicuspis

Family: Chenopodiaceae

GENERAL DESCRIPTION: Limestone bindii is an erect, short-lived perennial herb which grows to 25 cm. The branches are densely woolly and bear greyish green foliage covered with silky, white hairs. The fleshy, hairy leaves are up to 1 cm long, linear and round in cross-section. The flowers are insignificant, but numerous, and at fruiting appear to replace the leaves on the lower half of the stem. The seed is held in a woolly fruiting body about 4 mm across. It bears two red spines each about 5 mm long. They spread obliquely away from each other almost at right angles. Limestone bindii tends to grade into other bindii species, but can generally be distinguished by the colour and attitude of the spines.

HABITAT: Limestone bindii grows in calcareous areas in association with bluebushes, saltbushes and especially pearl bluebush.

Limestone bindii, about 20 cm high

FORAGE VALUE: It is unattractive to stock due to its spiny nature and it is eaten only as seedlings or in poor seasons. It contains up to 14 per cent crude protein and 10 per cent salt.

INDICATOR VALUE: Limestone bindii appears to increase under heavy grazing at the expense of more palatable bindiis and in large numbers indicates declining range condition.

REHABILITATION: Limestone bindii is only chosen in the colonising mix if the topsoil used to cover slopes is highly alkaline. If the topsoil is neutral or acid, other species are preferred.

Fruit have two sturdy curved spines, about actual size

Spear-fruit copperburr

Sclerolaena patenticuspis

Family: Chenopodiaceae

GENERAL DESCRIPTION: Spear-fruit copperburr is an herbaceous dwarf shrub growing to about 25 cm. The leaves are silky grey, succulent, cylindrical and 5–10 mm long. The fleshy flowers are very small, but produce a seed case 2–3 mm long bearing two divergent, twisting spines at the top. The spines are 5–8 mm long. The seed cases are often reddish on maturity. The seed cases vary a great deal and in some variants only one spine is present.

HABITAT: Spear-fruit copperburr is found in degraded saltbush and bluebush communities.

Spear-fruit copperburr, about 90 cm across

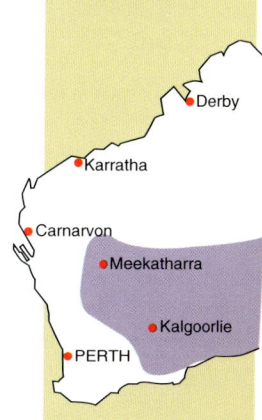

FORAGE VALUE: It is only eaten when green if no other feed is available.

INDICATOR VALUE: When present in large numbers and when it replaces other more palatable perennial plants, this species indicates poor range condition.

REHABILITATION: Spear-fruit copperburr can be used as a component of the 'bush litter' applied to areas undergoing rehabilitation. Whole plants should be collected in order to harvest the fruits.

Leaves and seed cases with spines, about one and a half times actual size

Sturt pea

Swainsona formosa

Family: Papilionaceae

GENERAL DESCRIPTION: Sturt pea is a prostrate annual or short-lived perennial creeper, growing to 4 m in diameter and to about 50 cm in height. The stems and leaves are covered in silky hairs. The leaves consist of nine to 15 spear-shaped leaflets, 1–3 cm long. The large pea-like flowers are red with a black or dark red centre and are up to 10 cm long. Groups of up to six are held in a circle at the end of short, upright stalks. The flowers produce swollen pods, up to 9 cm long and 1 cm broad, containing many black seeds 3 mm long. Albino or semi-albino pink forms of the plant are found in the Goldfields.

Sturt pea, about 1.2 m across

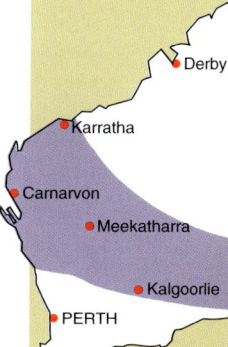

HABITAT: Sturt pea grows in areas receiving extra water, such as creek margins, or those with better soil types, such as lake margins. It grows on a wide range of alkaline soils, but not on acid soils. In the Pilbara it favours alkaline basaltic soils.

FORAGE VALUE: It is highly palatable to stock.

INDICATOR VALUE: It has no indicator value.

NOTE: It is a very popular garden plant. The seeds germinate readily after sanding, nicking or after hot water treatment, but establishment is a problem due to damping off. Once established, it can grow into a huge mat providing a spectacular show.

Flowers, about half actual size

Purple vetch

Swainsona pterostylis

Family: Papilionaceae

GENERAL DESCRIPTION: Purple vetch is a spreading annual with purple flowers. It grows to about 30 cm. In good winter seasons it can reach up to 1 m in diameter. The foliage is light green and each leaf consists of about eight pairs of leaflets. The tip of each leaflet is indented. The pea-like flowers are 9–10 mm long and up to 30 are held on spikes. The pods are up to 9 mm long and 7 mm wide. They have a wrinkled exterior and a deep fissure between the two halves.

HABITAT: Purple vetch is mainly found in locations that receive extra water such as creeks, floodplains, roadsides and the upper margins of salt lake systems. It favours neutral to alkaline soils.

Purple vetch, about 50 cm across

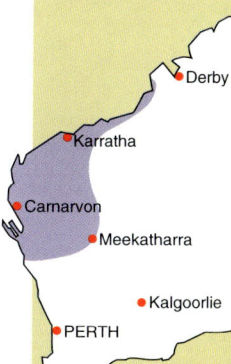

FORAGE VALUE: It is an excellent forage, with protein levels of 20 per cent having been recorded. It is highly palatable to stock.

INDICATOR VALUE: Its presence in dense stands indicates good seasonal conditions.

REHABILITATION: Purple vetch can be gathered easily in good seasons. It stores well and can be included in mixes applied to sites which are slightly saline and where extra water may be accumulated after rain. In mining areas it will be found most commonly on watercourses leading to lake systems.

Flowers and pods, about half actual size; pods becoming more wrinkled with age down the stem

Red spinach

Trianthema triquetra

Family: Aizoaceae

GENERAL DESCRIPTION: Red spinach is a dense, prostrate annual which grows to 10 cm high and 50 cm wide. It is usually dark green in colour, but turns reddish when mature. The whole plant is succulent with leaves which are round in cross-section, 1–2 mm in diameter and up to 1 cm long. The flowers are white and very small and are held in the leaf axils with the fruits.

HABITAT: Red spinach is found in semi-saline disturbed areas that are devoid of dense perennial vegetation and is common in areas which have been cultivated.

Red spinach, about 50 cm across

FORAGE VALUE: It appears to be rarely eaten by sheep or cattle.

INDICATOR VALUE: In dense stands it indicates semi-saline rangelands in poor range condition.

NOTE: It is a primary coloniser of degraded rangelands and is valued for its land stabilising capabilities.

REHABILITATION: Red spinach could be used in saline environments which receive run-on water. Seed is difficult to collect. Whole plants should be gathered when they are red in colour and spread directly onto the prepared surface as a means of distributing seed. It is unlikely that the plants could be stored successfully as they are very succulent.

Leaves and flowers, about twice actual size

Velleia

Velleia rosea

Family: Goodeniaceae

GENERAL DESCRIPTION: Velleia is a prostrate, winter-growing, annual herb with large, showy, pink flowers. The leaves are broad with blunt tips and have large teeth along the margins. The flowers are a pinkish white with broad petals, the whole flower being about 15 mm long. The seeds are 4–5 mm in diameter inclusive of a broad wing and are held in a capsule.

Velleia, about 50 cm across

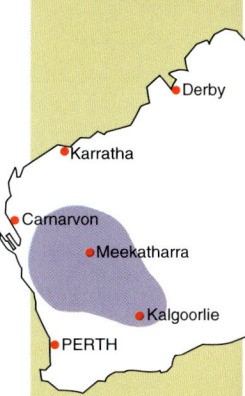

HABITAT: Velleia grows on hardpan plains beneath mulga. It also grows on shallow sandplains with bowgada. In good seasons it produces a dense mat of growth with a spectacular show of flowers.

FORAGE VALUE: It is not eaten by stock.

INDICATOR VALUE: It has no indicator value.

Everlasting daisies

Angianthus milnei
Brachycome ciliocarpa
Cephalipterum drummondii
Podolepis canescens
Rhodanthe chlorocephala subsp. *splendida*
Rhodanthe floribunda
Schoenia cassiniana
Waitzia acuminata

Family: Asteraceae

GENERAL DESCRIPTION: The everlastings are herbaceous annuals with showy papery flowers which range in diameter from 1 to 10 cm and in colour from pink and yellow to white. They germinate only on winter rain. A fall of 25 mm or more in May is all that is required for germination. They flower from June to August, and later blow away.

HABITAT: Everlastings grow in a wide range of habitats. They are most prevalent on loams and sandy clay loams beneath mulga and bowgada.

FORAGE VALUE: They are generally considered to be of good feed value. In most years they become dry in late spring and early summer and do not contribute much to diet from this point on.

Angianthus milnei, *about 20 cm high*

Brachycome ciliocarpa, *about 20 cm high*

Cephalipterum drummondii— *yellow form, about 25 cm high*

Cephalipterum drummondii— *white form, about 20 cm high*

Podolepis canescens, *about 20 cm high*

Rhodanthe floribunda, *about 10 cm high*

Schoenia cassiniana, *about 20 cm high*

Waitzia acuminata, *about 30 cm high*

Section II

Annual and perennial grasses

Long greybeard grass

Amphipogon caricinus

Family: Poaceae

GENERAL DESCRIPTION: Long greybeard grass is an erect perennial grass forming dense tussocks which are 20–60 cm high and up to 10 cm in diameter. The stems are wiry and support narrow, stiffly pointed leaves up to 30 cm long and a dense flower head 2–4 cm long and up to 5 mm wide. The young flower heads are initially purple, but turn yellow on maturity. Each seed carries five 1 cm long bristles which are covered in dense, grey hairs. At maturity the plant forms into a tight bundle tapering towards the top and is readily apparent among the spinifex with which it grows.

Long greybeard grass, young and growing vigorously, about 30 cm high

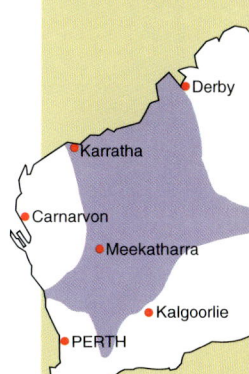

HABITAT: Long greybeard grass is restricted to spinifex sandplains where it can be found in dense populations during the post-fire regeneration process. As the spinifex assumes dominance, this grass becomes 'dormant' and less visible.

FORAGE VALUE: Although it is more attractive to stock than associated hard spinifex is, long greybeard grass is never consumed eagerly by stock and is only slightly grazed.

INDICATOR VALUE: Long greybeard grass has no indicator value.

Seed head, about two-thirds actual size

Wind grass   Kerosene grass

Aristida contorta

Family: Poaceae

GENERAL DESCRIPTION: Wind grass is an annual or weakly perennial grass that grows to 30 cm and forms compact, rounded tufts which produce numerous, sharp-pointed seeds. The leaves are short and narrow and roll inwards, curling as the plant dries off to its characteristic whitish and twisted appearance. Numerous reddish purple seeds are produced on the upper half of the plant. The long, thin seed has a sharp tip. It is surmounted by a prominent, twisted column about 3 cm long, which branches into three slender arms. On ripening, the seeds, columns and their branches fall to the ground in tangled masses.

HABITAT: Wind grass grows in almost every rangeland environment, avoiding only the clays and cracking clays.

Wind grass, about 20 cm high

FORAGE VALUE: It is the most widespread of the grasses and is readily eaten by stock, especially when it is green. Crude protein is then as high as 10 per cent. Stock will graze dry wind grass before it deteriorates following rain. Wind grass rapidly 'greens up' after small amounts of rain.

INDICATOR VALUE: Wind grass has no indicator value.

NOTE: The large bulk of seed produced by masses of wind grass plants in some years can be a severe hazard to sheep as the seeds mat together beneath the neck where they prevent the animal from drinking, or mat to form hobbles on the forelegs.

REHABILITATION: Wind grass is only suitable as a colonising plant on slopes which are neutral or acid in reaction. It is unlikely to be successful if the slopes are saline or alkaline.

Sharp-tipped seeds, about actual size

WA buffel grass

Cenchrus ciliaris

Family: Poaceae

GENERAL DESCRIPTION: WA buffel grass is a strong tussock grass growing to 50 cm. The tussocks are extremely dense at the base and produce prominently bent stems which often branch 20 cm or so above the ground. The stems are thick and tough at the base but become slender at their apex. The leaves are 7–30 cm long, 2–8 mm wide and taper gradually to a fine point. The flower heads form a single terminal cylindrical spike, 1.5–2.5 cm long and 0.5–1 cm wide. The individual spikelets in the spike are 4–5 mm in diameter and have soft bristles which characterise the plant. They consist of a small seed inside a capsule that is covered in stiff hairs. The seed heads are initially black or purple but turn white-yellow on maturity.

WA buffel grass, about 50 cm across

HABITAT: It grows on alluvial plains, on creek banks, and in areas of disturbance, such as roadsides, in frost-free environments. It does not tolerate highly acid soils and is most frequently found on alkaline soils and in association with limestone or alkaline rock breakdown products in rocky country.

FORAGE VALUE: When actively growing it can have a protein content of up to 16 per cent and digestibility of 70 per cent, but when dry, its protein content declines to 2 per cent. It is drought tolerant and responds to small amounts of rain. WA buffel grass is occasionally toxic to stock as it may contain large amounts of oxalates when mature.

NOTE: Birdwood grass (*C. setiger*) is a closely related species that is very similar to WA buffel grass vegetatively, though it has a knotty root base. The species can be separated by the seed heads. Birdwood grass seed heads are the same size as those of WA buffel grass, but have hard, tough bristles instead of the hairs on the spikelets.

Birdwood grass seed heads, about two-thirds actual size

WA buffel grass seed heads, about actual size

Ribbon grass Weeping grass, Golden beard grass

Chrysopogon fallax

Family: Poaceae

GENERAL DESCRIPTION: Ribbon grass is a strong, perennial, tussock-forming grass with leaves that droop to the ground. It has dense, fibrous butts which are heavily covered with soft, silky hairs. The tussocks are up to 30 cm across. The leaves mostly originate from below ground level, and are up to 25 cm long, 4–5 mm wide. The flowering stem is up to 1 m tall, upright, has small leaves and supports a loose head of purplish to golden brown seed heads 20 cm long and 10 cm wide. The individual spikelets are about 1 cm long and three are borne at the end of each branch. The individual seeds are 1 cm long and each has a tuft of golden brown hairs around its spear-like base as well as a reddish brown, twisted, stout hair up to 4 cm long.

Ribbon grass, about 1.2 m high

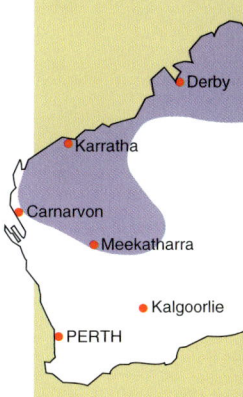

HABITAT: Ribbon grass occurs in areas subject to inundation such as creek banks, floodplains and swampy areas. It grows with soft spinifex, silky browntop, coolabah and mulga.

FORAGE VALUE: It is a valuable forage species and is extremely resistant to drought and heavy grazing. It is palatable when green, but once dry it is unacceptable to stock. Crude protein varies from 9 per cent when green to 3 per cent when dry, while digestibility varies from 60 per cent when green to 35 per cent when dry.

INDICATOR VALUE: It is an indicator of good range condition. Ribbon grass was once an important component of the pasture in the watercourses and creeklines of the Gascoyne and the northern Murchison, but it has been largely removed by overuse.

Flowering seed head, about half actual size

Lemon-scented grass   Scented oil grass

Cymbopogon ambiguus

Family: Poaceae

GENERAL DESCRIPTION: Lemon-scented grass is an erect, tussocky perennial grass which grows to 70 cm. The tussocks are up to 10 cm across. The leaves are bluish-green when young, becoming reddish at maturity, and are up to 30 cm long and about 2 mm wide. They arise at the base of the plant and taper to a fine point. They become very curled when dry. The seed heads consist of numerous tight spikes of seeds borne on the top of stout stems, up to 70 cm tall. Each seed is surrounded by a mass of straight, white hairs, up to 5 mm long, from which protrudes a 1 cm long, fine, slightly bent bristle or awn. The leaves and stems of this grass are strongly lemon-scented when crushed.

Lemon-scented grass, about 60 cm high, on granite

HABITAT: Lemon-scented grass grows on rocky hills, exposed granite and on roadsides which have shallow loam or clay soils. It also grows along the banks of creeks in stony uplands. It grows with a wide variety of species.

FORAGE VALUE: It is rarely eaten by stock.

INDICATOR VALUE: It has no known indicator value.

Seed head, about actual size

White top   Common wallaby grass

Danthonia caespitosa

Family: Poaceae

GENERAL DESCRIPTION: White top is a low, perennial grass generally about 30–40 cm high, though it may reach 60 cm. It forms a dense tussock up to 12 cm across. The leaf sheaths around the stems are strongly ribbed and may be hairy or free of hairs. The leaf blades spread gently away from the stems and are soft, bright green and slightly inrolled. The flower heads are a pale gold and are usually short, being about 4–5 cm long. The glumes which support each of the flowering spikelets remain intact after the seeds fall.

White top, about 30 cm high

HABITAT: White top grows on clay soils as well as on sandy clay loams. It appears to avoid sandy soils. It grows with saltbush and bluebush in less salty situations and sparsely in mulga and curara communities. It is most important on the Nullarbor Plain on clayey soils but is not significant elsewhere.

FORAGE VALUE: It is a valuable, palatable grass that makes most of its growth after winter rains, when temperatures begin to rise.

INDICATOR VALUE: The indicator value is unknown as it tolerates grazing very well.

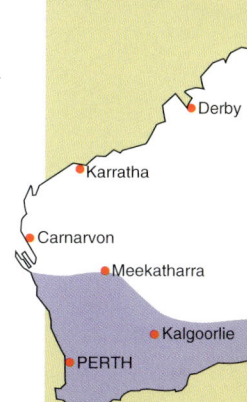

Seed heads, about actual size

Limestone grass

Enneapogon caerulescens

Family: Poaceae

GENERAL DESCRIPTION: Limestone grass is a short-lived, upright perennial grass, 15–40 cm high. It has short, narrow leaves which are produced from the base and joints of the stems. They are up to 5 cm long and 2 mm wide. The seeds are borne in a single, short, compact, dark grey head up to 2 cm long on the top of upright, slender stems up to 40 cm tall. The seed heads are covered with numerous fine hairs.

HABITAT: Limestone grass occurs in any areas with alkaline soils such as calcrete platforms adjoining rivers. It also grows on loamy and clayey soils on basaltic hills and rises. It grows with silver saltbush, sago bush, tall saltbush and bloodbush.

Limestone grass, about 20 cm high

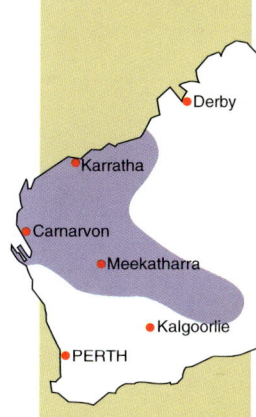

FORAGE VALUE: It is an excellent forage but does not provide a large bulk of feed. It disappears in periods of prolonged moisture stress, only reappearing after heavy, drought-breaking rains.

INDICATOR VALUE: Its presence indicates good or fair range condition.

Seed heads, about twice actual size

Windmill grass

Enteropogon acicularis

Family: Poaceae

GENERAL DESCRIPTION: Windmill grass is a weakly tussocking, perennial grass with dark brown windmill-like flower heads growing to 50 cm. The dark, bluish green leaves are up to 20 cm long and 4 mm wide, finely tapering to a sharp point. The dry leaves become curled and twisted and are a pale straw colour. The flower heads are borne erect on short stems about 10 cm above the foliage and consist of two to 14 large arms, up to 13 cm long, that all originate from a common point and give the appearance of a windmill head. The seeds lie flat on the arms and quickly

Windmill grass, about 20 cm high

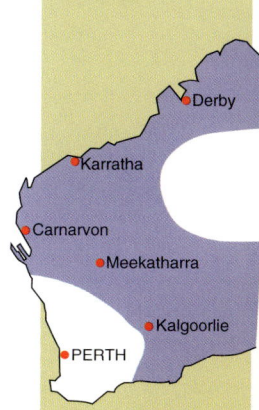

fall off when mature. They are up to 5 mm long and 1 mm wide and possess a 10–15 mm long, fine, brown bristle or awn.

Windmill grass is not a robust perennial grass; the leaves and stems do not persist with the onset of dry conditions.

HABITAT: Windmill grass grows on soils varying from sands to clays but generally there are only one or two plants within the protection of a shrub or tree.

FORAGE VALUE: The young growth is palatable to stock, but mature growth seems unpalatable.

INDICATOR VALUE: Its presence indicates good to fair range condition. It disappears under heavy grazing, but is stimulated by moderate grazing. It was an important pasture species on alluvial plains, but now tends to be rare.

Murchison red grass   Love grass, Red grass

Eragrostis dielsii

Family: Poaceae

GENERAL DESCRIPTION: Murchison red grass is a prostrate or erect annual or short-lived perennial, very variable in habit but generally growing in tufts no more than 20 cm high and 30 cm wide. The leaves are 1–2 cm long, rigid and rough along their length. The stems are shiny and terminate in large, spreading or upright, brownish red, curved, spike-like heads. They break up to yield very small, waxy seeds. In good seasonal conditions the grass forms a dense mat. The massed seed heads present a brilliant splash of colour.

Murchison red grass, about 30 cm across

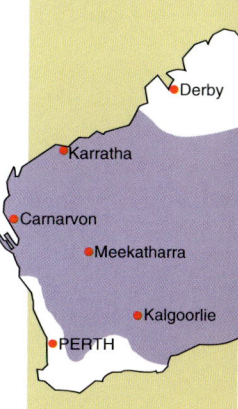

HABITAT: Murchison red grass is found in most habitats in the mulga lands, being most prevalent in alluvial run-on or watercourse country which is seasonally moist. It also occurs beneath breakaways and volunteers readily on rangeland regeneration sites.

FORAGE VALUE: It is highly sought after by sheep and kangaroos when green or dry. If it is spoilt by rain it turns grey-black and is unacceptable.

INDICATOR VALUE: It has no indicator value.

Seed heads, about one and a half times actual size

Wire wanderrie grass   Woolly butt

Eragrostis eriopoda

Family: Poaceae

GENERAL DESCRIPTION: Wire wanderrie is a robust, tussocky grass, growing to 60 cm. It has thick, woolly butts and hairy roots. The leaves are up to 10 cm long and flat when actively growing, but they roll up when the plant dries off and then become round, thin and pointed. They spread stiffly away from the stems. The stems are fine and very wiry. The seed head is 12–15 cm long with one or two spreading branches. The remaining branches are so short that the individual spikelets appear to be attached to the stem. Wire wanderrie grass tussocks can reach a great age, often more than 20 years. They tend to break up into apparently separate plants, or may alternatively develop a ring-form tussock.

Tussocks of wire wanderrie grass, about 30 cm high

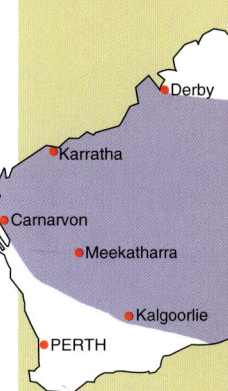

HABITAT: Wire wanderrie grass occurs on the sandy banks of broad outwash plains and on sandplains with moderately deep (0.4–1 m) acidic, sandy loams or sands. It generally grows with Wilcox bush, broad-leaved wanderrie grass, soft wanderrie grass, mulga and wanderrie wattle. In the Pilbara it is also found with soft spinifex.

FORAGE VALUE: It appears to be eaten only when green as it is very fibrous when dry. Stock, particularly sheep, actively select individual plants against others when grazing wire wanderrie pastures and consume the new growth on the grazed plants in preference to those mature plants which are adjacent. When green, wire wanderrie has a crude protein of 10 per cent which drops to 4 per cent when dry. It appears to be less palatable than broad-leaved and soft wanderrie grasses.

INDICATOR VALUE: This grass can withstand the harshest of conditions. It increases in density in holding paddocks and burnt areas due to the elimination of less durable species. Because it is extremely resistant to drought and grazing, it has little indicator value. In the Pilbara it is an early and valued volunteer after burning.

Seed heads, about one and a third times actual size

Creeping wanderrie grass

Eragrostis lanipes

Family: Poaceae

GENERAL DESCRIPTION: Creeping wanderrie grass is a tussocky, short-lived perennial grass which can grow to 50 cm, but more commonly it spreads its stems across the ground. It has a bulbous base which is slightly woolly, but it is not as dense as wire wanderrie grass. The leaves are about 10 cm long and 2 mm wide and are tightly rolled around the stem, except at the end where they may extend away for about 10 cm. The seed heads are a loose, open panicle of quite small spikelets which are a brownish purple when young.

Creeping wanderrie grass, about 30 cm high

HABITAT: Creeping wanderrie grass grows with other wanderrie grasses on sandy banks and plains where the soils are red, loamy sands over 30 cm deep. It is also common on sand dunes around salt lakes.

FORAGE VALUE: It is a preferred forage plant and contains up to 16 per cent crude protein. It does not survive moisture stress well and is therefore not as valuable as the other wanderrie grasses.

INDICATOR VALUE: Although it is preferentially eaten by stock, it is of limited value as an indicator of range condition since it does not survive medium-term moisture stress.

Seed heads, about actual size

Neverfail

Eragrostis setifolia

Family: Poaceae

GENERAL DESCRIPTION: Neverfail is an upright, slender grass which grows to 50 cm in dense, round tussocks. It has a thickened base which is slightly woolly. The bright green leaves branch from the stems and not from below the ground. They are fine, not exceeding 1 mm wide, gently curled and up to 10 cm long. There are many long, slender, upright stems supporting the seed heads which consist of dense, spike-like panicles.

Neverfail, about 30 cm high

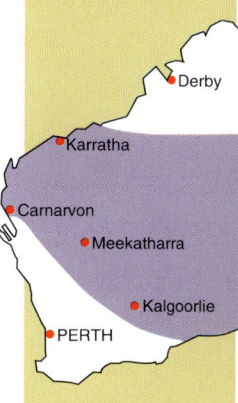

HABITAT: Neverfail grows in swamps, crabholes and drainage floors with clay soils that are frequently slightly saline. It grows with swamp grass, Roebourne Plains grass, creeping cassia, swamp saltbush, curara and river red gum.

FORAGE VALUE: Neverfail is readily eaten by stock, even when dry. It is resistant to grazing and responds quickly to rainfall, though it is slow to respond when the temperatures are low.

INDICATOR VALUE: Plant numbers decrease under very heavy stocking and its presence indicates good range condition.

Seed head, about one and a half times actual size

Roebourne Plains grass Knotty-butt neverfail

Eragrostis xerophila

Family: Poaceae

GENERAL DESCRIPTION: Roebourne Plains grass is a dense, tussocky, perennial grass with a butt up to 30 cm across. The stem bases within the butt are bulbous, scaly and most commonly woolly, but some may be free of this dense, white covering. The stems are thin and the leaves extend away from them about halfway up. The leaf blades are rather short and hairless, but soft and a mid-green colour. The seed heads are purplish and occur in a loose panicle about 10 cm long. The seed is produced in herringbone-shaped spikelets which are up to 1 cm long and 3 mm wide.

HABITAT: Roebourne Plains grass grows on clays and heavy loam soils along the Roebourne Plains and in crabholes and linear gilgais with

Roebourne Plains grass, about 30 cm across

cracking clay or heavy clay soils in the interior. It is found with neverfail and bundle bundle.

FORAGE VALUE: It is a highly drought resistant and palatable perennial grass which is tolerant of grazing. Stock prefer annual species growing with it, but Roebourne Plains grass provides the bulk of the feed on offer in periods when annuals fail.

INDICATOR VALUE: It should be a common component of crabholes and gilgais in the Pilbara. If it is absent, range condition may have deteriorated. An absence of the plant on its preferred soil type on the Roebourne Plains indicates poor range condition.

Seed heads, about half actual size

False wanderrie grass

Eriachne aristidea

Family: Poaceae

GENERAL DESCRIPTION: False wanderrie grass is an annual or very short-lived perennial, tufted grass that never forms a robust tussock. It grows to about 45 cm. The base is sometimes thickened, but is never woolly or hairy. The stems are slender and usually bent near the ground. The leaves are flat, 5 mm wide, and extend away from the stems for up to 15 cm. The seed head is loose and open, bearing solitary spikelets and resembling broad-leaved wanderrie grass. In false wanderrie grass, however, each spikelet has bristles and hairs attached. When young the spikelets are purple, but after the seed is shed the pale yellow outer husks remain attached to the plant stem.

False wanderrie grass, about 30 cm high

HABITAT: False wanderrie grass grows on sandy banks and on sandplains beneath bowgada, wanyu and mulga.

FORAGE VALUE: False wanderrie grass is not palatable and is usually trampled or blown away. In New South Wales it is reported to be readily grazed when green, but rapidly loses its attractiveness as it matures.

INDICATOR VALUE: The presence of abundant false wanderrie grass indicates poor range condition. When plentiful it will have replaced more desirable species.

A seed head, about actual size

Swamp grass
Eriachne benthamii

Family: Poaceae

GENERAL DESCRIPTION: Swamp grass is a strong, upright, tussocky perennial grass with erect stems which are occasionally branched. It grows to about 90 cm, though it is commonly less. The densely hairy butt is about 15 cm in diameter. The leaves are a bluish green and have a grey, waxy covering. The leaf blade is about 15 cm long and 3–5 mm wide. The flower head is a loose, brown-green panicle, up to 10 cm long and 2–3.5 mm wide. The individual spikelets are 7–10 mm long and are covered by dense hairs in the lower half. The upper half terminates in a recurved bristle.

Swamp grass, about 50 cm high

HABITAT: Swamp grass is found in areas which are frequently inundated. It is common on drainage swamps and in claypans. It will also grow on cracking clay plains as well as on non-cracking clay soils.

FORAGE VALUE: It is generally not eaten by stock except in unusual situations such as holding paddocks. It remains green into the dry season because of the favourable habitat.

INDICATOR VALUE: It is an increaser species in cracking clay grasslands. Populations increase under heavy grazing at the expense of weeping and barley Mitchell grasses and Roebourne Plains grass.

Seed heads, about actual size

Claypan grass

Eriachne flaccida

Family: Poaceae

GENERAL DESCRIPTION: Claypan grass is a dense, smooth, tussock-forming perennial which grows to 40 cm. It has a thickened, very slightly hairy base. The leaves are narrow and rolled tightly around the stems, though towards maturity they may extend away about 15 cm. The seed heads are compact panicles up to 8 cm long and do not extend much above the vegetative growth. The outer husks remain attached to the stem after the seed falls.

Claypan grass, about 50 cm high

HABITAT: Claypan grass is found in claypans, crabholes and swampy areas. It is also found growing in the clay beds of some rivers.

FORAGE VALUE: It is palatable, being particularly favoured by cattle. Mature plants have a crude protein content of 6 per cent, but younger plants and selected portions of older plants could be expected to be more nutritious. It is resistant to grazing and to drought and is a valuable though restricted source of feed.

INDICATOR VALUE: The presence of claypan grass indicates good range condition as plant numbers decrease under excessive grazing.

A seed head, about one and a half times actual size

Buck wanderrie grass

Eriachne helmsii

Family: Poaceae

GENERAL DESCRIPTION: Buck wanderrie grass is a harsh, robust, tussock-forming perennial grass which grows to 60 cm. It has a bulbous, woolly base which can be 60 cm across. The stems bend close to the ground before they ascend and thus give the plant a very open appearance. The leaves are flat and curl around the stem for most of their length, but finally extend away in the last 10 cm. Residual leaf material from previous years of growth often persists around the lower parts of the plant, giving it a coarse appearance. The seed head is an open head or panicle and up to 10 cm long. It is not unlike broad-leaved wanderrie grass and produces only a few, pale, straw-coloured spikelets about 5 mm long in each panicle. After seed drop the outer husks of each spikelet remain attached to the stem.

Buck wanderrie grass, about 60 cm high

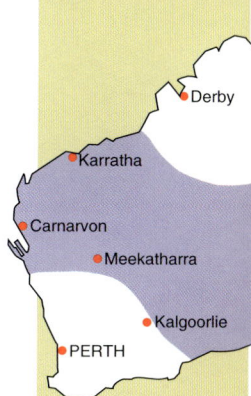

HABITAT: Buck wanderrie grass grows on sand banks and plains in red loamy sands with mulga, Wilcox bush and tall saltbush.

FORAGE VALUE: Buck wanderrie grass is generally unpalatable and is seldom grazed.

INDICATOR VALUE: Buck wanderrie grass is rarely an abundant species though in some situations it may dominate a pasture. Even so, it is a good indicator of poor range condition in the Murchison and the Gascoyne. When it is the only perennial grass present on sand bank and sandplain vegetation in these regions, it indicates fair to poor range condition.

Immature and mature seed heads, about actual size

Silky browntop

Eulalia fulva

Family: Poaceae

GENERAL DESCRIPTION: Silky browntop is a dense, tussocky, long-lived perennial grass which grows commonly to 1 m but it can reach 1.5 m and be 60 cm in diameter at the base. The leaf blades are flat and about 20 cm long and 5 mm wide. They taper gradually to a fine point and turn a distinctive red-brown when dry. The seed heads are borne on erect stems and consist of two to four distinctive, brown, silky spikes 5–10 cm long, usually of unequal length.

Silky browntop, about 80 cm high

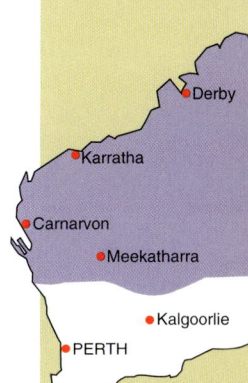

HABITAT: Silky browntop favours wetter areas in the shrublands, particularly the margins of creeks and water holes and broad, flat watercourses which have deeper soils than usual. There is some suggestion that it was more common on the alluvial drainage plains of the Murchison, in particular at the time of settlement, but the plant could have been confused in the vegetative state with kangaroo grass.

FORAGE VALUE: It is a drought resistant plant of moderate palatability when green and is unpalatable when dry.

INDICATOR VALUE: Because of its restricted occurrence, silky browntop is of limited value as an indicator plant. Should it be observed colonising seasonally wetter areas, it could indicate improving range condition.

A seed head, about actual size

Broad-leaved wanderrie grass Bandicoot grass

Monachather paradoxa

Family: Poaceae

GENERAL DESCRIPTION: Broad-leaved wanderrie grass is a vigorous, tussock-forming perennial grass which grows to 60 cm. It has a thickened, often slightly woolly base which can be up to 25 cm across. The stems are numerous and unbranched. The leaves are harsh to the touch and up to 15 cm long and 5 mm wide. The flower head is a loose and open panicle, up to 15 cm long. The seeds drop shortly after they mature and leave the papery, pale ivory, outer husk behind.

HABITAT: Broad-leaved wanderrie grass is most common on sandy banks and sandplains where it is associated with wire wanderrie and buck wanderrie grasses. It can also be found on shallower soils.

Broad-leaved wanderrie grass, about 30 cm high

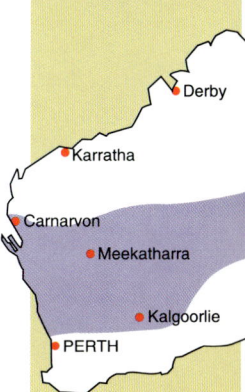

FORAGE VALUE: Broad-leaved wanderrie grass has a crude protein content as high as 15 per cent though this declines with maturity. It is relished by stock and is one of the more important grasses of wanderrie vegetation. It has the capacity to respond vigorously to both summer and winter rain, but germinates only in the summer.

INDICATOR VALUE: In the river catchments of the Murchison and the Gascoyne, broad-leaved wanderrie grass is an indicator of good range condition if it is found dominating a pasture. In the Goldfields it behaves opportunistically in response to rain and cannot be used as an indicator of range condition there. In this region wire wanderrie grass appears to be the dominant, drought-resisting perennial grass, with broad-leaved wanderrie volunteering in good seasons.

Mature and immature seed heads, about one third actual size

Oat-eared spinifex   Soft spinifex

Plectrachne schinzii

Family: Poaceae

GENERAL DESCRIPTION: Oat-eared spinifex forms dense hummocks which are up to 1 m wide and 50 cm high. It rarely forms the vegetative rings which are characteristic of true spinifex species. It has dense, green, semi-weeping leaves with resinous stems. The leaves are up to 25 cm long, flattish and pointed at the tips. The seed heads rise on stems 25–50 cm above the hummock. The seed head is about 12 cm long and has a feather-like, open appearance due to the long, sturdy bristle on each seed. The combined seed and bristle are about 2 cm long.

HABITAT: Oat-eared spinifex grows on sand dunes with deep, red, sandy soils. It grows with grevilleas, sand wattle, other spinifex species, wire wanderrie grass and a range of herbaceous shrubs.

Oat-eared spinifex, with seed heads, about 1 m high

FORAGE VALUE: After it is burned, the fresh growth of oat-eared spinifex is eaten by stock. At this stage it mainly consists of young leaves. The crude protein varies from 10 per cent when young to 3 per cent when old, and digestibility varies from 35 per cent when young to 25 per cent when old. It persists well during droughts and is a good drought reserve.

INDICATOR VALUE: Its presence on deep, sandy soil indicates good to fair range condition.

NOTE: Although oat-eared spinifex resembles spinifex vegetatively, it can be distinguished by its feathery seed head. In spinifex, the seed head is much shorter and dense rather than open.

Part of a seed head, about one and a half times actual size

Rats tail couch   River couch

Sporobolus mitchellii

Family: Poaceae

GENERAL DESCRIPTION: Rats tail couch is a mat-forming perennial grass 20–45 cm high. The bases of the stems are hairless and are thickened as a result of the covering of papery leaves. The above-ground stems are thin and wiry and can spread for up to 3 m. The leaves are 3–10 cm long, broad and are tapered to a fine point. The flower heads consist of a tan-coloured, narrow spike, 6–12 cm long and 4–6 mm wide.

HABITAT: Rats tail couch is found in areas subject to inundation which remain wet for protracted periods. It is common around river pools, swamps, springs and coastal flats. It is not found on cracking clay soils.

Rats tail couch, about 50 cm across

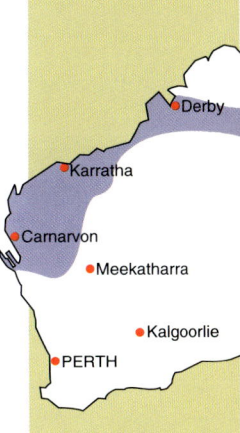

FORAGE VALUE: Sheep and cattle graze this species when green.

INDICATOR VALUE: This is a decreaser species and large populations of this grass indicate good range condition.

Seed heads, about half actual size

Silver speargrass   Feather speargrass

Stipa elegantissima

Family: Poaceae

GENERAL DESCRIPTION: Silver speargrass is a large, tussocky grass growing to 1 m, but is commonly about 50–75 cm in height. The tussocks are up to 30 cm across. The stems are stiff, thin and branched at the nodes. The leaves are a dull, deep green and are narrow and inrolled. They lack any hairs and extend away from the stem for about 10 cm. The flowering head is very conspicuous. It is a loose panicle with the individual spikelets covered in grey to purplish feathery hairs. Each tussock bears several seed heads which together, at flowering, top the plant with a gossamer-like covering. The feathery awns are quite fragile and are easily broken up. The seeds soon fall but they do not present the problems to stock that those of other species of *Stipa* do.

Silver speargrass tussock, about 75 cm high, supported by shrubs

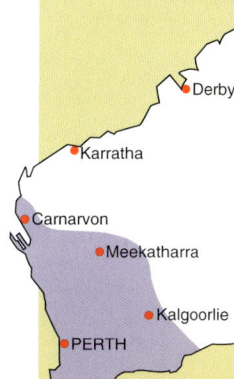

HABITAT: Silver speargrass is most common in the less saline saltbush communities and favours sites which receive additional water. It is also found in the alkaline soils beneath York gums. It grows with silver saltbush, sage and sago bush.

FORAGE VALUE: It is readily eaten by stock.

INDICATOR VALUE: Silver speargrass is an indicator of good range condition when it is found growing in saltbush pastures. It is not common in heavily grazed pastures, where it is found growing within other plants which afford it some protection. This characteristic has given rise to the mistaken impression that it depends upon other plants for support. In ungrazed areas, silver speargrass can be found growing in isolation from other plants.

Single seed head, about actual size

Speargrass

Stipa scabra

Family: Poaceae

GENERAL DESCRIPTION: Speargrass is a tufted perennial grass with stems that grow 30–80 cm. The grass blade is linear, curled round the stem at the basal part, but standing out from it at the free end. The flower spike is 10–20 cm long and up to 3 cm broad. There are many long bristles, each 5–7 cm long, one to each seed, which become strongly recurved when ripe. They are spirally coiled where attached. The name 'speargrass'

Speargrass, about 40 cm high

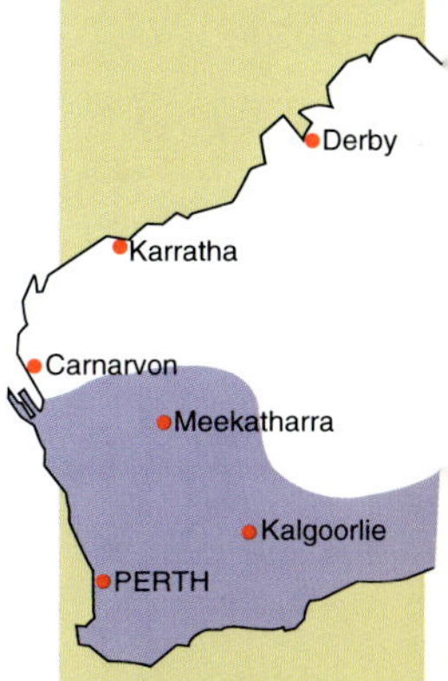

is due to the hard, sharp base of the seed when it is shed. This can be harmful to the skin of sheep if it becomes tangled in the wool.

HABITAT: Speargrass is restricted to alkaline loams and grows with bluebush, limestone wattle and broom bush.

FORAGE VALUE: It is readily eaten when young and green, but tends to be ignored by stock when it is dry and harsh. It contains up to 6 per cent crude protein and 5 per cent salt.

INDICATOR VALUE: Speargrass has no known indicator value.

Speargrass seed, about one and a half times actual size

Kangaroo grass

Themeda triandra

Family: Poaceae

GENERAL DESCRIPTION: Kangaroo grass is a tussock-forming, leafy, perennial grass which grows to 1.5 m. The tussock may be 25 cm across. The leaf blades are narrow, bright green and hairless or with a few scattered, straight hairs. They turn brown to red towards maturity. The same species in South Africa is called red grass. The seed-bearing stems

Kangaroo grass, about 1.3 m high

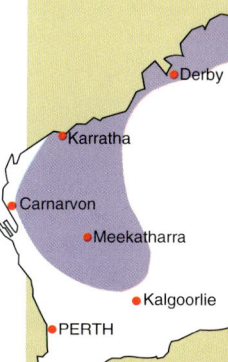

rise well above the leaves and the seeds are carried in an open panicle up to 25 cm long. The seed heads are tinged a brown-red at maturity. Each spikelet in the head is about 2 cm across. The awns from the spikelets are slightly hairy, twice bent, loosely twisted and are 5–7 cm long.

HABITAT: Kangaroo grass is usually found in favoured locations such as in creeks and along well-watered creek margins with mulga, river gum and black mulga. It is thought that it was more widespread at settlement but, being highly palatable and not resistant to heavy grazing, it has become rare.

FORAGE VALUE: It is an excellent fodder grass though becoming coarser and less attractive with maturity. Overseas work with the species shows that it will disappear if it is grazed continuously.

INDICATOR VALUE: The presence of kangaroo grass on creek margins and on alluvial plains indicates excellent range condition.

Seed heads, about actual size. Note bent and twisted awn on each spikelet

Soft wanderrie grass   Mulga grass, Barbed wire grass

Thyridolepis multiculmis

Family: Poaceae

GENERAL DESCRIPTION: Soft wanderrie grass is a tussock-forming perennial grass which grows to 60 cm from a densely packed base. The stems are numerous, erect and straw-coloured when dry. The leaves are bright green, about 5 mm wide and 5 cm long, extending horizontally away from the stems and ending in a firm point. They are soft and pliable to the touch while still growing, but grow markedly stiffer and more sharply pointed when dry. The seed head is a dense panicle up to 5 cm long and the outer parts carry stiff bristles which are sticky to the touch. It responds vegetatively to both winter and summer rain and will seed in the winter as well as in the summer.

Soft wanderrie grass, about 30 cm high

HABITAT: Soft wanderrie grass grows on deep sandplain soils which receive run-on water, beneath trees in mulga groves and in mulga thickets in the lower Murchison. It is rarely found where the soil is less than 30 cm deep and is not common on wanderrie banks.

FORAGE VALUE: It is a palatable grass containing up to 16 per cent crude protein when green, but the protein content falls markedly with maturity.

INDICATOR VALUE: It is an indicator of good range condition and populations decrease under heavy grazing.

A seed head, about one and a half times actual size

Hard spinifex

Triodia basedowii

Family: Poaceae

GENERAL DESCRIPTION: Hard spinifex is a harsh, prickly, tussocky perennial grass which forms a low hummock. The sharply pointed leaves are bluish green in colour and covered with a faint, white bloom. They are 5–25 cm long and 2 mm broad. The seeding stems are erect, rising to 60–70 cm, and may be branched. The seed head is 8–12 cm long. Each of the seed heads contains six to eight spikelets.

Hard spinifex, a mature tussock in ring form, about 2 m across

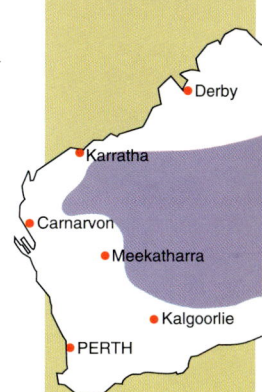

HABITAT: Hard spinifex is found on sandplains with mallees and low shrubs.

FORAGE VALUE: It has no forage value.

INDICATOR VALUE: It has no indicator value for range condition.

NOTE: Hard spinifex is a general term used to describe *Triodia* species which are especially prickly and unpalatable.

The hard spinifex group includes *Triodia lanigera* and *T. wiseana* in the Pilbara and *T. intermedia* in the Kimberley as well as other species in both regions.

Soft spinifex is the common name given to *Triodia pungens*.

Seed heads, about actual size

Soft spinifex

Triodia pungens

Family: Poaceae

GENERAL DESCRIPTION: Soft spinifex is a perennial hummock grass growing to 1 m and up to 1.5 m wide and with highly resinous stems and leaves. It does not usually form rings. All the leaves are held in the lower parts of the stems and form the basis of the hummock. The stems sometimes produce runners. The leaves are up to 30 cm long. When young or in humid conditions the leaves are flat, but they curl up with the onset of dry conditions. The leaves also are slightly curled on the vertical axis. The stems produce an upright flower head which consists of an open, pyramid-like panicle of spikelets, 3–10 cm long.

HABITAT: Soft spinifex is found in a wide range of habitats from basalt hills to cracking clay plains.

FORAGE VALUE: It is eaten by sheep and cattle during periods of forage shortages. Young soft spinifex is almost totally edible, with a protein

Soft spinifex, young tussock, about 1.2 m across

content of between 7 and 11 per cent. Five-year-old soft spinifex has a protein content of 4–5 per cent. It is an excellent drought reserve.

INDICATOR VALUE: On granitic plains, large populations indicate good range condition. It has the ability to invade cracking clay environments when the vigour of the tussock grasses has been reduced through overgrazing. In this situation large populations indicate poor range condition.

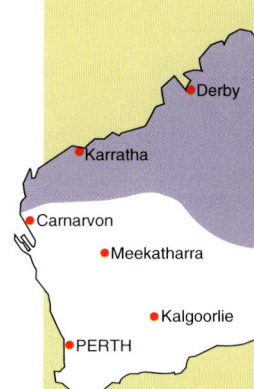

Seed heads, with spikelets in a narrow panicle, about actual size

SECTION III

Shrubs less than 3 m high

ACACIA GROUP

Poverty bush

Acacia translucens

Family: Mimosaceae

GENERAL DESCRIPTION: Poverty bush is a low, spreading, flat-topped shrub up to 1 m high and 1.5 m wide. It has hairless, almost oval, shiny, dark green leaves, 1–2 cm long and about 7 mm wide. The flower heads are yellow, globular and up to 5 mm in diameter. The seed pods are held prominently erect above the foliage and do not droop below it as in most *Acacia* species. They are up to 9 cm long and 8 mm wide and are flattish and woody when dry.

Poverty bush, about 70 cm high. Note the flat top of the shrub

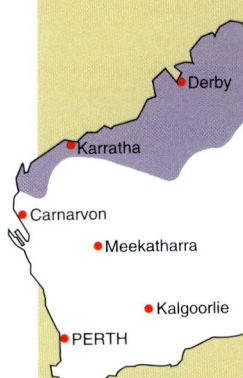

HABITAT: Poverty bush occurs on spinifex plains with medium depth sands and loams and grows with soft spinifex, hard spinifex, *Acacia bivenosa* and *Eucalyptus setosa*.

FORAGE VALUE: Poverty bush is unpalatable and has no forage value.

INDICATOR VALUE: When abundant, it is an indicator of spinifex country in poor range condition. Spinifex country which is managed incorrectly by repeated winter burning will become dominated by this plant and will produce little forage for stock. It can be suppressed by summer burning and deferred grazing, which will restore vigour to the spinifex and associated grasses. Subsequent management using the appropriate burning and stocking strategies will prevent its re-establishment in quantity.

Distinctive upright pods and round-ended leaves, about actual size

Swamp saltbush   River saltbush

Atriplex amnicola

Family: Chenopodiaceae

GENERAL DESCRIPTION: Swamp saltbush is a spreading, multi-branched shrub up to 1.5 m high. The branches spread across the ground. Under favourable conditions they may layer and take root. The leaves are usually spear-shaped, though some ecotypes have margins which are not lobed. The leaves are a bluish green and up to 2.5 cm long and 4 mm wide. They are covered with a grey, mealy sheen of fine silvery hairs. Male and female flowers are borne on separate plants. The male flowers are gathered into globular, purplish to green spikes at the ends of the branches. The female flowers cluster in the axils between leaves and stems and also in short spikes at the ends of the branches. They are profuse and on maturity each forms a stout, woody seed case, about 5 mm square, containing a single seed.

HABITAT: Swamp saltbush grows in areas which are regularly inundated such as floodplains, river channels and lake fringes. It is also found in the wetter parts of sluggish upland drainages. It grows with samphire, silver saltbush and saline mulla mulla.

Swamp saltbush, about 2 m across and 1 m high

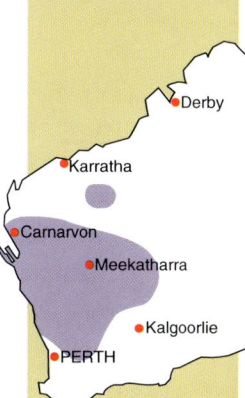

FORAGE VALUE: Swamp saltbush is palatable and contains about 10 per cent crude protein. It is fairly drought tolerant, persists well into very dry periods and responds vigorously to small showers of rain.

INDICATOR VALUE: If pastures contain vigorous swamp saltbush plants in a range of ages and sizes, they may be said to be in good range condition. Care must be taken in assessing range condition on the basis of vigorous seedling storms. Such seedlings can easily die if proper grazing management is not practised.

REHABILITATION: Swamp saltbush demands a moister environment than other saltbushes. Persistence could therefore be a problem where extra water cannot be conserved locally. It will colonise well in the less arid sites as it establishes easily and seeds profusely. Several cultivars are available. 'Bencubbin' has proved the most successful to date.

Seed cases and foliage, about half actual size

Silver saltbush

Atriplex bunburyana

Family: Chenopodiaceae

GENERAL DESCRIPTION: Silver saltbush is an erect, open, many-branched shrub which grows to about 1.5 m. The branches are rigid and slender. The leaves are about 10 mm long, generally elliptical in shape and with a smooth outline. They have a thin, bluish, scaly covering. There are separate male and female plants. The male flowers occur in small, globular clusters and are borne in short panicles. The female flowers also occur in small open panicles which arise at the ends of the branches and along the stems. At maturity they appear as thin, papery seed cases, almost heart-shaped and resembling two leaves joined together. They contain a single seed and often bear two horn or antler-like structures on each side near the base.

HABITAT: It is found in the less saline parts of saltbush pastures and in the tributary drainages leading to major drainage features. It is common on floodplains and sand dunes surrounding salt lakes throughout its range. It also grows on some basaltic hill slopes in the Goldfields and the Murchison.

Silver saltbush, about 50 cm high

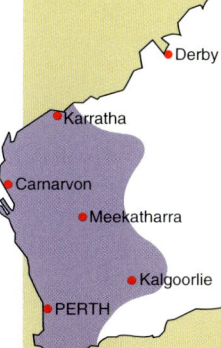

FORAGE VALUE: It is a palatable plant and is eaten at all stages of its growth if fresh water is available. It contains up to 12 per cent crude protein and 15 per cent salt. When plants of the herbage layer are available it tends not to be grazed, but is eaten as soon as these dry off. It retains its leaves well into periods of moisture stress, but it is not as drought resistant as sago bush, which grows with it in many places. Silver saltbush shoots again rapidly with small falls of rain and is an important component of the pasture.

INDICATOR VALUE: Silver saltbush is an indicator of good range condition. When present as vigorous well-grown plants over a range of ages and sizes, the vegetation is in good condition.

REHABILITATION: Silver saltbush has been used successfully in mine site reclamation on the less saline sites.

Seed cases and foliage, about actual size

Old man saltbush

Atriplex nummularia

Family: Chenopodiaceae

GENERAL DESCRIPTION: Old man saltbush is a bushy shrub which grows to 2 m high and 3 m wide. The smaller woody branches are brittle, but the principal stems are stout and robust. The leaves are elliptical to ovate, about 2 cm across, and have a slightly irregular outline. They have a white, scaly covering which gives them a greyish appearance. Male and female flowers are carried on separate plants. The male flowers are produced in globular heads at the ends of the branches. The female flowers are produced prolifically at the ends of the branches and along the stems. As with all saltbushes the flowers have no petals. Large leaf-like structures develop around the seed after it is fertilised. The seed cases are about 1 cm across and turn yellow on maturity. On occasions they may be borne so freely as to cause the branches to break.

HABITAT: Old man saltbush grows on undulating limestone plains and on lake margins with alkaline loams and clay soils. It grows with silver saltbush, bladder saltbush and pearl bluebush and is common beneath eucalypts in the Goldfields gum belt.

Old man saltbush, about 1.7 m high

FORAGE VALUE: Old man saltbush varies in palatability, being eaten in some instances and avoided in others in the same paddock. In many cases a high salt content in the leaves could be responsible for its unattractiveness to stock. It contains about 14 per cent crude fibre and has a digestibility of about 52 per cent. Provided it is palatable, it is a valuable component of a pasture.

INDICATOR VALUE: In the Goldfields, populations of old man saltbush increase in heavily grazed paddocks, where it tends to replace bladder saltbush. In these situations it indicates fair or poor range condition. In the eastern Murchison and Gascoyne, populations decline under heavy grazing. Healthy populations here indicate good range condition.

REHABILITATION: Old man saltbush has been used extensively in mine dump reclamation. It tolerates alkaline and saline conditions. It seeds prolifically in September and October and is easily collected. Not all seed cases contain seeds and a check should be made on the percentage of seed before it is sown. It is usually sown at 3 kg of seed-filled cases per hectare, but it is better included with a range of other species to give diversity to the stand.

Seed cases and foliage, about half actual size

Male flowers and foliage, about half actual size

Kidney saltbush

Atriplex stipitata

Family: Chenopodiaceae

GENERAL DESCRIPTION: Kidney saltbush is an erect, open, multi-branched shrub growing to about 1 m high, though commonly less. It resembles silver saltbush and may be confused with it. The leaves are up to 2 cm long and may be elliptic and pointed at the end, or else rounded. They have the bluish grey sheen on the upper and lower surfaces that is common in saltbushes. Male and female flowers grow on separate plants. The male flowers occur in globular clusters at the ends of the branches. The female flowers occur on long, thin spikes. They are succeeded by the bracts which are kidney-shaped, rising from a short stem. The bracts are 5–10 mm wide and they lack appendages. The single brown seed is found inside the bracts.

HABITAT: Kidney saltbush is found on calcareous and slightly saline soils beneath salmon gums and gimlets and Clelands blackbutt. Other plants found with it include broom bush.

Kidney saltbush, about 60 cm high

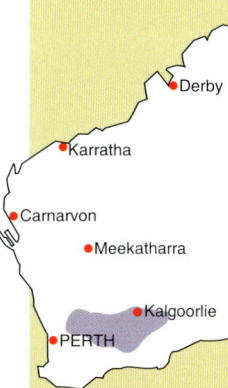

FORAGE VALUE: The leaves of kidney saltbush contain a bitter principle which apparently makes the plant unattractive to stock.

INDICATOR VALUE: As it is unpalatable, it has no indicator value. Pastures containing large communities of kidney saltbush appear to be in first-class condition and capable of supporting large numbers of stock. In fact the shrubs have no grazing value.

REHABILITATION: The seed of kidney saltbush is readily harvested in the Goldfields. If stored under air-dry conditions, seed should keep for 12 months or more. It may be mixed with other saltbushes and bluebushes to seed waste dumps and tailings dam walls. Its unpalatability confers an establishment advantage not shared by palatable species.

Male flowers, and distinctive leaves, about half size

Bladder saltbush

Atriplex vesicaria

Family: Chenopodiaceae

GENERAL DESCRIPTION: Bladder saltbush is a low-growing shrub which reaches 80 cm. The stems, which are produced freely, are very brittle and snap easily underfoot. The leaves are greyish, about 2 cm long, and have a grey, mealy covering. They are oval and arranged alternately along the stem. The margins are smooth and never notched. Leaf size reflects the current moisture conditions: those produced in periods of good rainfall are large and fleshy. There are separate male and female plants. Male plants produce globular heads of very small flowers at the ends of the youngest branches. Female flowers are very small and are produced along the stems. The seeds are enclosed in large, leaf-like structures about 5 mm across, which carry a spongy appendage on each side.

Bladder saltbush, about 40 cm high

HABITAT: Bladder saltbush is found on alkaline loams and clays beneath gums, on alluvial duplex soils and on salt lake margins and saline drainages.

FORAGE VALUE: Bladder saltbush is a valuable source of feed for stock. In some instances high salt content limits its usefulness since it can contain up to 15 per cent salt. It contains up to 12 per cent of crude protein. The leaves tend to drop with moisture stress, but the plant responds quickly to rainfall.

INDICATOR VALUE: Bladder saltbush is a reliable indicator of good range condition. Populations regenerate well on the Nullarbor, but are slow to recover in the Murchison and the Gascoyne. Management of country to promote bladder saltbush will have to be opportunistic and flexible to take advantage of the good rainfall years.

REHABILITATION: Bladder saltbush is a useful species to include in mine site rehabilitation in the Goldfields. Seed should be collected locally and included with other species and 'bush litter'.

Seed case and foliage, about two and a half times actual size

Cassia group

Cassias are woody perennials found in most habitats in the dry interior. They can be 3 m high, but are commonly less. Most have a characteristic leaf form. The leaves consist of a central rib which supports a number of leaflets occurring in opposite pairs along its length. The number of pairs varies from one in desert cassia, to seven or more in straight leaf cassia.

All species have attractive buttercup-yellow flowers which have five separate petals. Long, flattened pods develop from the flowers. These rattle when dry and led to the general name of 'Rattlepods' for the cassia group.

Some members of the group, such as creeping cassia, are highly palatable to stock whilst others, such as grey cassia, are generally unpalatable and become woody weeds in overgrazed situations.

There are about 20 species of cassia in Western Australian arid shrublands, of which nine are described here.

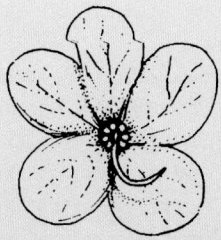

A cassia flower. All cassia flowers are very similar and cannot be used to differentiate species

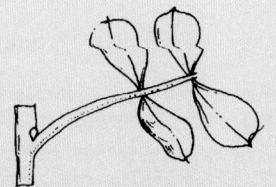

Bloodbush leaf, *C. oligophylla*

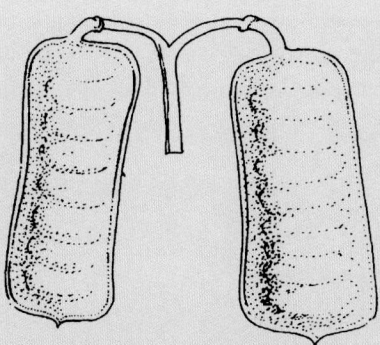

Pods of crinkled cassia, *C. helmsii*

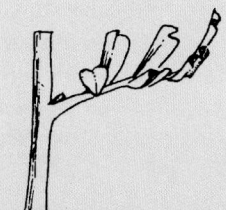

Crinkled cassia leaf, *C. helmsii*

Desert cassia leaf, *C. nemophila*

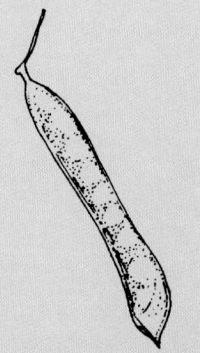

Pods of white cassia, *C. luerssenii*

White cassia leaf, *C. luerssenii*

The genus *Cassia* has been revised to *Senna* in recent botanical literature. The older convention of *Cassia* is followed here.

Flowers, leaves and pods of some cassias

CASSIA GROUP

Silver cassia

Cassia artemisioides

Family: Caesalpiniaceae

GENERAL DESCRIPTION: Silver cassia is a small, bushy shrub which grows to 2 m. The leaves are covered with a silvery grey bloom which is easily rubbed off. They are up to 6 cm long and bear three to eight pairs of leaflets which are 2–4 cm long and about 1 mm across. The flowers are bright yellow and occur in groups of up to ten. The pods are flat, thin and brown in colour and contain dark green to black seeds.

HABITAT: Silver cassia grows in a wide range of communities and is found on rocky ridges, on lower slopes and on alluvial plains with mulga and curara. It also grows with soft and hard spinifex on sands and rocky soils.

Silver cassia, about 1.2 m high

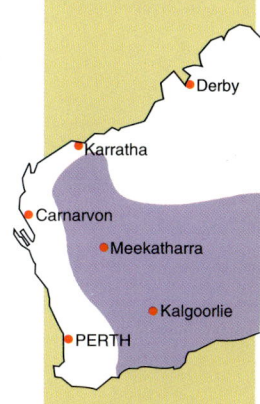

FORAGE VALUE: Although the protein content is high, it is not readily eaten by stock.

INDICATOR VALUE: Silver cassia is a rapid growing species capable of colonising an area but then disappearing as quickly since it is short-lived. It has no indicator value.

REHABILITATION: Silver cassia germinates rapidly from seed if a pre-soaking treatment is given. Seed supplies are easily obtained. It may be included with other shrubby species on dump slopes which are non-saline. It tolerates high alkalinity.

Pods and distinctive narrow, silver-grey leaflets, about half actual size

CASSIA GROUP

Green cassia

Cassia chatelainiana

Family: Caesalpiniaceae

GENERAL DESCRIPTION: Green cassia is a branched shrub which grows to 3 m. The branches and stems lack the well-developed felty covering which is common on some other *Cassia* species. It is a long-lived shrub with a robust stem. The leaves are usually dark green and variable in shape and size. There are usually three or more pairs of leaflets with a gland between the bottom pair. The flowers are a deep yellow and are more sparse than those of other *Cassia* species. They are succeeded by long pods which turn brown at maturity.

Green cassia, about 1.2 m high

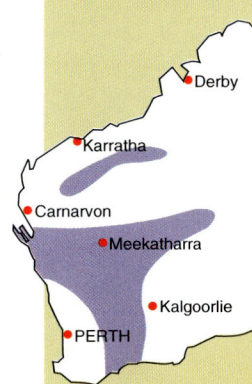

HABITAT: Green cassia is most common in non-saline pastures of wanderrie shrublands and floodplains where it grows on shallow loamy sands. It is associated with mulga, turpentine bush, tall sida, tall saltbush and flannel bush.

FORAGE VALUE: Green cassia is a valuable forage shrub and is grazed closely by stock. Older plants assume an umbrella shape showing a well-defined grazing line; this is an aid to field identification. It retains its leaves well into periods of moisture stress, but in extreme droughts may lose all its leaves and die.

INDICATOR VALUE: Pastures with vigorous green cassia plants of all ages are in good range condition.

Pods and foliage, about actual size

Flowers and foliage, about actual size

CASSIA GROUP

Straight leaf cassia

Cassia desolata

Family: Caesalpiniaceae

GENERAL DESCRIPTION: Straight leaf cassia forms part of a complex taxonomic group in which hybridisation is common. It is not surprising, therefore, that it is an extremely variable shrub. It is 1–2 m high and the branches may be covered with a slight down. The leaves are divided into two to seven pairs of leaflets. In some specimens the leaflets may be up to 1 cm across and in others they are much narrower. They are usually covered with a dense down, are rounded or pointed at the tips, and are about 1–2 cm long. In some specimens the down may be lacking altogether or be much reduced. The flowers are a very showy buttercup yellow. They are borne in groups of four or five on the branchlets. The flowers are succeeded by pods up to 7 cm long. These darken and become brittle with age.

HABITAT: Straight leaf cassia is found throughout the shrublands, avoiding only the spinifex communities and the lowermost or very saline parts of the saltbush pastures. It is commonly found on wanderrie interbanks and the fringes of wanderrie sand banks.

Straight leaf cassia, about 1 m high

FORAGE VALUE: Straight leaf cassia varies in its palatability. Some varieties are eaten, principally those with little downy covering. For the most part, however, the plant is not grazed. Protein contents of about 13 per cent are recorded.

INDICATOR VALUE: If the plant is identified as an unpalatable form, straight leaf cassia can be a useful indicator plant. When it is found in large numbers it indicates poor range condition.

Flowers, immature pod and foliage, about half actual size

CASSIA GROUP

Creeping cassia

Cassia hamersleyensis

Family: Caesalpiniaceae

GENERAL DESCRIPTION: Creeping cassia is a low, spreading shrub up to 40 cm high and 60 cm wide. The leaves consist of two to four pairs of almost triangular leaflets each up to 2 cm long and with indented tips. It flowers in response to rain. The yellow flowers are 2 cm wide and are held in bunches of about six on a common stalk up to 6 cm long. Numerous flattened, black seeds, up to 6 mm long and 3 mm wide, are borne in flat, brown pods up to 6 cm long and 12 mm wide.

HABITAT: Creeping cassia occurs in drainage depressions, around crabholes and on cracking clay soils and grows with Roebourne Plains grass and neverfail.

FORAGE VALUE: It is a useful forage, providing green material when associated grasses are dry. It does not persist well during dry periods, but quickly produces forage after rain.

INDICATOR VALUE: It is a decreaser species and its presence indicates good range condition.

Creeping cassia, about 30 cm high

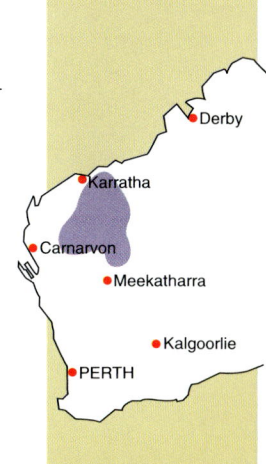

Flowers and almost mature pods, about actual size

CASSIA GROUP

Crinkled cassia   Blunt-leaf cassia, Grey cassia

Cassia helmsii

Family: Caesalpiniaceae

GENERAL DESCRIPTION: Crinkled cassia is an open shrub of marked grey appearance which grows to 2 m, but is more commonly about 1 m high. The upper branches are cloaked with a dense covering of short, grey hairs, but this is lacking on the lower parts which have a fissured bark. The leaves are 3–4 cm long and consist of up to five pairs of leaflets which have blunt-tipped and rolled-over points, giving their edges a crinkled or wavy appearance. The flowers are buttercup yellow and grow in groups of up to ten, on new wood only. The pods, which follow, are flat, oblong, about 6 cm long, and dry to a dark brown at maturity.

HABITAT: Crinkled cassia grows in most rangelands in the shrub zone, but it avoids the spinifex communities and highly saline areas.

Crinkled cassia, about 70 cm high

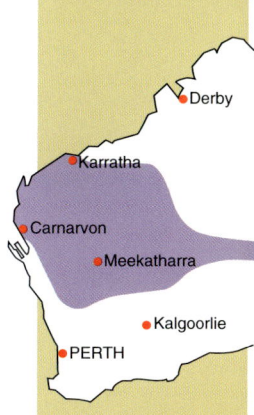

FORAGE VALUE: It is normally not palatable, though stock may be driven by feed shortage to graze it sparingly. Its protein content is about 12 per cent.

INDICATOR VALUE: Crinkled cassia is not a desirable species. When it dominates a community to the exclusion of other species it indicates poor range condition. It readily colonises sandy and sandy loam soils and has formed almost closed communities on some sandy banks in the Gascoyne and the Murchison, rendering them unproductive. It is also often found in the sandy beds of streams in disturbed and degraded rangelands.

Flowers, foliage and immature pods, about half actual size

CASSIA GROUP

White cassia

Cassia luerssenii

Family: Caesalpiniaceae

GENERAL DESCRIPTION: White cassia is an upright, greyish green shrub up to 1.5 m high. The leaves are covered in a white, waxy bloom and consist of seven pairs of narrow leaflets each up to 15 mm long and 3 mm wide. The flowers are bright yellow and up to 2 cm wide. They are held in groups of four at the end of a stalk which is about 3 cm long. They are succeeded by flat, obtuse pods which have a white wax covering initially, but they lose it with maturity and become a deep brown.

White cassia, about 1 m high

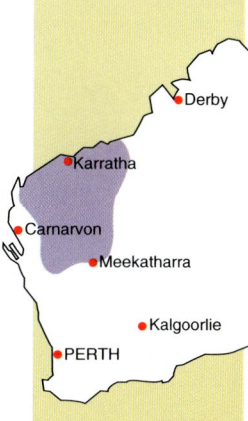

HABITAT: White cassia is found on rocky hills and on plains with shallow, acidic loams above rock or hardpan. It grows with mulga, wind grass and various poverty bushes.

FORAGE VALUE: It is rarely eaten and has no forage value.

INDICATOR VALUE: It has no known indicator value.

Flowers, immature pods and foliage, about three-quarters actual size

CASSIA GROUP

Desert cassia   Punty bush

Cassia nemophila

Family: Caesalpiniaceae

GENERAL DESCRIPTION: Desert cassia is an erect, bushy shrub which can grow to about 2 m, but is more commonly 1 m high. The stems and branches lack any downy covering. The leaves vary but are usually 5–7 cm long and consist of one or two pairs of narrow or even rounded, hairless, bright green leaflets. The leaflets may be 3 cm long. One form of desert cassia has leaves up to 5 mm across, but it is rare and restricted to the run-off areas around rocky outcrops. The flowers are borne in profusion in groups of three to ten. They are a bright yellow and are followed by pods which are up to 8 cm long and up to 1 cm wide. These turn brown at maturity.

Desert cassia, about 1.3 m high

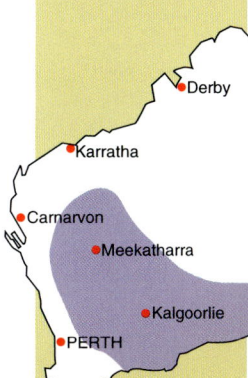

HABITAT: The fine-leaved, dominant form of desert cassia is common in opaline areas and alkaline soils.

FORAGE VALUE: Desert cassia does not appear to be palatable though the seed pods are eaten.

INDICATOR VALUE: When desert cassia is present in dense stands, it indicates poor range condition. In New South Wales it is one of the shrubs which have invaded the grasslands and open woodlands of the Western Land Division, causing them to be much less productive. There is evidence to suggest that this situation may be arising in the Goldfields. This is a matter of some concern, since it is a comparatively long-lived species which will be difficult to dislodge by grazing or resting alone. There is some suggestion that it can be controlled by fire.

In Western Australia the adoption of grazing practices which favour more desirable species will be the principal tool used to prevent the spread of desert cassia. In cases of severe infestation, burning may be the only feasible method of control.

Flowers, foliage and immature pods, about half actual size

CASSIA GROUP

Bloodbush   Buttercups

Cassia oligophylla

Family: Caesalpiniaceae

GENERAL DESCRIPTION: Bloodbush is an upright shrub growing up to 1.5 m. It has multiple stems which branch from the base. The leaves are about 10 cm long and consist of up to three pairs of leaflets. The leaflets are up to 3.5 cm long and 1.5 cm wide, often rounded at the ends, sometimes slightly hairy, and always covered with a white, waxy bloom. The flowers are bright yellow, 2 cm wide, and are borne in bunches of about six on stalks 4 cm long. The flattish, black seeds are 7 mm long and 4 mm wide, and are held in papery brown pods up to 7 cm long and 2 cm wide.

HABITAT: Bloodbush grows in a wide range of habitats including floodplains, alkaline shale plains, limestone plains and crabhole plains. It appears to be restricted to soils which are neutral or alkaline.

Bloodbush, about 60 cm high

FORAGE VALUE: It is eaten by all stock, especially horses, when more attractive forage has been depleted. It stains the saliva blood-red when it is chewed, hence its common name. It sheds its foliage at the onset of dry conditions and is not a good drought reserve.

INDICATOR VALUE: On floodplains it is an increaser species, replacing more desirable plants such as river saltbush and tall saltbush. It is a decreaser species on shale and limestone plains. It is relatively short-lived and succumbs easily to droughts and insect damage, but it recruits rapidly from seed. It is not a reliable indicator of range condition and must be used in conjunction with other key species to assess trends in condition.

Pods and foliage, about two-thirds actual size

CASSIA GROUP

Variable cassia

Cassia sturtii

Family: Caesalpiniaceae

GENERAL DESCRIPTION: Variable cassia is an upright, flat-topped shrub growing to 2 m. It has multiple stems arising from the base. The yellowish green leaves have four pairs of hairless leaflets 5–20 mm long and about 3 mm wide. They have a grey, waxy covering in most cases, but this may be absent and the shrub will appear bright green. The flowers are 15 mm wide and bright yellow. They occur in bunches of about five on a short, common stalk up to 2 cm long. Numerous flattened, black seeds up to 8 mm long and 4 mm wide are carried in golden brown pods, which are up to 8 cm long and 2 cm wide.

HABITAT: Variable cassia occurs on creeks and wanderrie banks with acid, sandy loam soils over hardpan and grows with mulga, Wilcox bush, wanderrie grasses and turpentine bush. It may occur infrequently in the less saline parts of saltbush pastures.

Variable cassia, about 1.3 m high

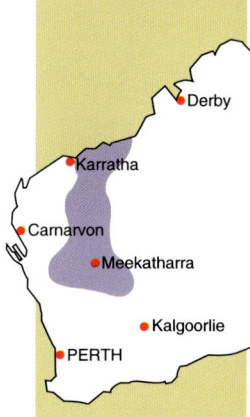

FORAGE VALUE: It is eaten by stock when annual feed is short. It tends to lose its leaves during prolonged dry periods and is, therefore, of limited value. It belongs to an extremely variable group in which there is great genetic diversity. Some ecotypes may not be grazed at all, while others are eagerly sought after by stock. Leaves contain about 13 per cent crude protein.

INDICATOR VALUE: It is a decreaser species. As it is rarely abundant, it is an unreliable indicator of range condition.

Flowers and foliage, about actual size

Swamp bluebush Northern bluebush

Chenopodium auricomum

Family: Chenopodiaceae

GENERAL DESCRIPTION: Swamp bluebush is an upright shrub with erect, branched stems and grows to about 2 m. The leaves are spear-shaped and up to 5 cm long. They have a dense, scaly, grey surface below and a scarcely hairy, blue-grey upper surface. The leaves have a distinctive fishy odour when crushed. The flowers are very small and occur in spikes or broad panicles at the ends of the branches. The fruiting body is small and contains a small, black seed.

HABITAT: Swamp bluebush is found in heavy soils including some cracking clays. It grows with neverfail and Roebourne Plains grasses.

Swamp bluebush, about 1 m high

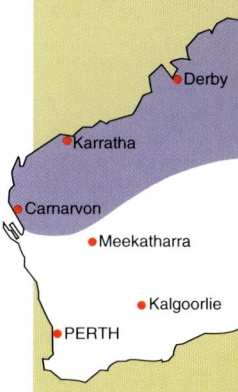

FORAGE VALUE: It is readily eaten by stock, particularly when the ground feed has dried.

INDICATOR VALUE: It is an indicator of good range condition. It is extremely tolerant of grazing. When pastures contain only a few heavily grazed plants of swamp bluebush, heavy overuse must be suspected.

Flowers and foliage, about actual size

Scrambling saltbush   Cottony saltbush

Chenopodium gaudichaudianum

Family: Chenopodiaceae

GENERAL DESCRIPTION: Scrambling saltbush grows to about 2 m, but is more commonly about 1 m high. The stems are fragile and slender, often with long internodes between the leaves. It supports itself only weakly and tends to rely upon other trees and shrubs. The leaves are alternate, rather thick and characteristically triangular or spearhead-shaped. They have a dense, hairy cover and are up to 2.5 cm long and 2 cm wide. The flowers are clustered at the ends of the branches. They are very small and are followed by small, orange or red berry-like fruits, 2–3 mm across. Gathered together in clusters, they resemble bunches of very small grapes. The seeds are very small, one only being found at the centre of each of the berries.

Scrambling saltbush, about 60 cm high

HABITAT: Scrambling saltbush grows on sands and clays in many different communities, but tends to avoid the open saltbush shrublands. It is frequently found beneath limestone wattle and mulga.

FORAGE VALUE: It is a palatable and useful source of forage.

INDICATOR VALUE: It is a good indicator of range condition. When mature robust plants are evident, the country is in good range condition. Seedlings grow freely after rain. Range condition should not be judged using numbers of plants alone, since the mortality of the seedlings, even under ideal conditions, can be very high.

REHABILITATION: Scrambling saltbush is a good coloniser of waste dump and tailings dam batters. As it seeds profusely, it is capable of spreading rapidly. It favours alkaline soils and is moderately salt tolerant. If seed is collected, it must be dried well and kept in a dry environment, as the fleshy covering will promote fungal infestations.

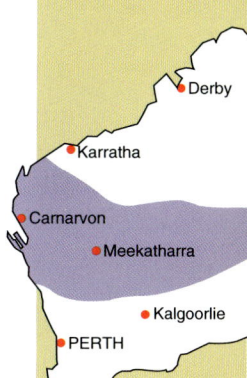

Mature and developing berries and foliage, about one and a half times actual size

Grey corchorus

Corchorus walcottii

Family: Tiliaceae

GENERAL DESCRIPTION: Grey corchorus is a low, open, sparsely branched, often straggly shrub, which can reach 1 m, but is commonly less. It is short-lived and never has a robust appearance. The stems and leaves are very woolly. The leaves are thick, broadly ovate to oblong, and have smoothly notched margins. The flowers are borne singly, or in small groups on short stems in the axils of leaf and stem. They are five-petalled, yellow and about 1 cm across. There are many stamens surrounding the ovary or undeveloped fruit. The small rounded seed case or fruit is about 5 mm across and contains five seeds.

HABITAT: Grey corchorus is found on sandy and sandy loam soils on sandplains and dunes. It may occur after fire in spinifex communities and is a common plant beneath acacias on sand dunes and sandy banks.

Grey corchorus, about 50 cm high

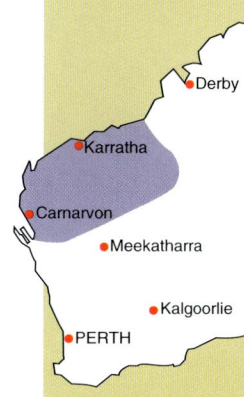

FORAGE VALUE: It does not appear to be browsed by stock and has no forage value. It is commonly the only low shrub left untouched on sand dunes and sand banks when all other low and accessible species have been consumed. It contains about 10 per cent crude protein.

INDICATOR VALUE: Grey corchorus has no known indicator value.

Flowers, fruit and foliage, about actual size

False bluebush

Cratystylis conocephala

Family: Asteraceae

GENERAL DESCRIPTION: False bluebush is a roundish, dense, bluish grey shrub which grows to 1.5 m. It strongly resembles pearl bluebush and is often confused with it. The bluish grey leaves and stems are covered in dense, short, grey hairs. The leaves are up to 8 mm long, flattened to oval in cross-section, and broaden towards the tips. They are arranged alternately and often profusely along the stem. The flowers are borne at the ends of the branchlets and have very small, white petals.

False bluebush, about 1.3 m high

HABITAT: It grows on alkaline sandy loam and loamy soils with western myall, salmon gum and bladder saltbush.

FORAGE VALUE: False bluebush is sometimes eaten by stock.

INDICATOR VALUE: It has no known indicator value.

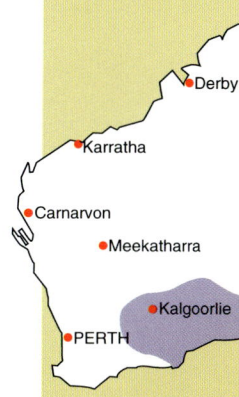

Flower buds about to open, about twice actual size

Sage   Australian sage

Cratystylis subspinescens

Family: Asteraceae

GENERAL DESCRIPTION: Sage is a many-branched, dense shrub up to 2 m high. The ends of the twigs form blunt thorns. The greyish green leaves have a pungent, sage-like smell when crushed and are smooth, hairless and up to 1 cm long and 3 mm wide. They are not succulent. Young shoots are very long and vigorous and often a pale, greenish yellow. When immature they are soft and flexible, but as they mature they become stiff and angular. The flowers are small, white and daisy-like and up to 5 mm across. The ripe seed heads dry out and open to expose feathery, thistle-like seeds.

HABITAT: Sage is common on heavy soils associated with river floodplains, inland salt lakes, crabholes and sluggish, saline, upland drainages. It grows with silver saltbush, fuchsia bush and narrow-leaved mulla mulla.

Sage, about 1.3 m high

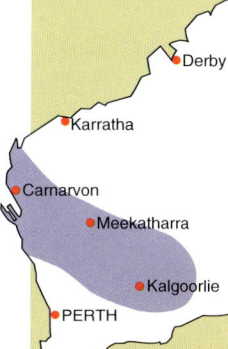

FORAGE VALUE: While it is not as palatable as the saltbushes with which it grows, it is eaten in preference to these plants when stock waters are saline. The leaves can contain up to 9 per cent crude protein. It is fairly drought resistant, but sheds its leaves under extreme drought conditions.

INDICATOR VALUE: Sage is very resistant to grazing and can indicate overuse when it is the dominant plant in salt lake and saline river vegetation. Under extremely heavy and continuous use, sage will disappear. It is an indicator of good range condition in the more restricted and harsh environments of saline hill pastures and on basaltic hills, but is slow to regenerate when overgrazed.

REHABILITATION: Although sage is a common plant of saline environments, it is not likely to be used in mine site rehabilitation since the seeds are difficult to harvest. Some plants may establish if twigs holding ripening seeds are laid on prepared batters.

Foliage and pointed side branches, about actual size

An open flower and foliage, about actual size

Bead hopbush   Wild hopbush

Dodonaea lobulata

Family: Sapindaceae

GENERAL DESCRIPTION: Bead hopbush is a multi-branched shrub which grows to 2 m. The leaves are bright green and sticky. They are arranged alternately along the stem and are up to 3 cm long and about 3 mm across. The edges of the leaves are marked with distinctive, often pointed, lobes. In bead hopbush there are separate male and female plants. The photograph shows the female plant with its brown fruits, which have three prominent, papery wings up to 1.5 cm wide. The small seed lies in the centre of the wings. The male flowers are small. Hopbush is so-named because of the similarity of its winged fruit to those of true hops used in brewing.

HABITAT: Bead hopbush grows on rocky hills and loams and tolerates shallow soils.

Bead hopbush, about 1.3 m high

FORAGE VALUE: It may be occasionally eaten by stock, but generally has no forage value.

INDICATOR VALUE: It has no indicator value.

REHABILITATION: Bead hopbush should be a component of seed mixes applied to waste dumps which are acid or only slightly alkaline. Dumps associated with lateritic deposits are particularly well suited. The plant will provide drought-resistant low cover and will augment the trees which are more commonly used.

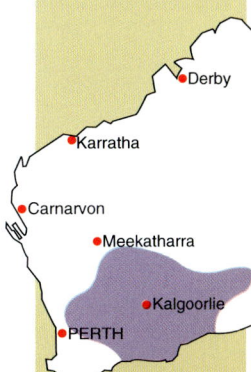

Lobed leaves, and brown, winged fruits, about three-quarters actual size

Sticky hopbush   Broad-leaf hopbush

Dodonaea viscosa

Family: Sapindaceae

GENERAL DESCRIPTION: Sticky hopbush is an upright shrub which grows to 3 m. It has reddish, sticky stems. The leaves are bright green and sticky, occurring alternately along the stems. They are up to 8 cm long and up to 1.5 mm broad and the edges may be slightly wavy. The leaf veins are very prominent. There are separate male and female plants. The male flowers are small and are gathered into short panicles. The fruits on female plants have three papery wings and may be red or purplish in colour.

Sticky hopbush, about 1.8 m high

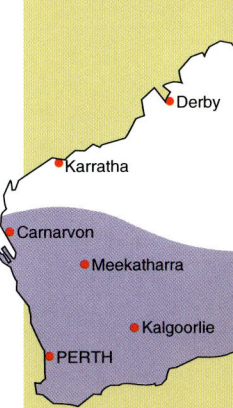

HABITAT: It is most common on rocky hillsides and with gravels. It grows beneath eucalypts and also in mulga communities, particularly in groves where it can be very dense beneath the tree canopy.

FORAGE VALUE: It is only sparingly eaten by stock.

INDICATOR VALUE: In Western Australia it has no recognised indicator value. In eastern Australia and in New South Wales in particular, it colonises disturbed and degraded areas and can indicate a deterioration in range condition there.

Three-winged, crimson fruits and narrow, sticky leaves, about half actual size

Ruby saltbush

Enchylaena tomentosa

Family: Chenopodiaceae

GENERAL DESCRIPTION: Ruby saltbush is an open, often straggling shrub, which grows to 1 m, but is commonly only half that height when grazed. It is variable in form and leaf colour. It has soft, brittle stems and succulent leaves. The leaves and stems are usually densely woolly. In rare cases the plant is hairless. The leaves are a greyish green and are arranged alternately along the stems. They are round in cross-section, 1–2 cm long and pointed at the tips. The flowers are small and inconspicuous and are found singly in the leaf axils. They are succeeded by prominent, very succulent, orange or red berries which turn black when dry.

HABITAT: Ruby saltbush is found in almost all communities in the shrublands. It grows on practically all soil types from sands to clays.

Ruby saltbush, about 40 cm high

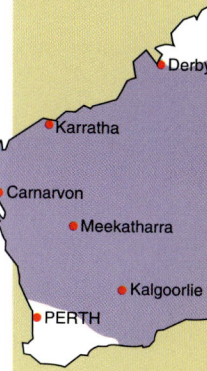

FORAGE VALUE: It is an important component of the diet of sheep and is readily eaten. It contains up to 14 per cent crude protein, 6 per cent salt and is 65 per cent digestible.

INDICATOR VALUE: An abundance of mature ruby saltbush indicates that an area is in good range condition. It is not, however, a reliable indicator of changes in range condition, since it produces seedlings freely after rain in much the same fashion as scrambling saltbush. The young seedlings often succumb through competition even though conditions are favourable. Range condition should only be assessed on the mature members of the population.

REHABILITATION: Small quantities of ruby saltbush should be included in seed mixes sown on non-saline to moderately saline batters. Seed is usually obtainable from commercial sources. If gathered locally, it should be dried before storing as the fleshy outer covering could lead to fungal infestations.

Fruit and foliage, about two-thirds actual size

Poverty and fuchsia bush group

Poverty and fuchsia bushes belong to the genus *Eremophila* and are common plants in the arid parts of Australia. They are woody species varying from low shrubs to small trees. They are found in all except the most saline situations. The group is characterised by tube-like flowers which range in colour from white through blue and green to pink, brown and red. They are very showy, and in some species the sepals behind the flower expand and become highly coloured after the petal tube drops.

The leaves of all species have a distinctive aromatic odour when crushed.

Some species are palatable to stock, but most are ignored or eaten only very sparingly.

There are about 212 species of Eremophila in Australia, of which about 180 are found in Western Australia and 160 in the arid shrublands. The Wiluna–Meekatharra area is the centre of diversity for this plant group. The 23 species described in this book are the most widespread members of the group.

Illustrations from the *Journal of Adelaide Botanic Gardens*, 1979 and 1980.

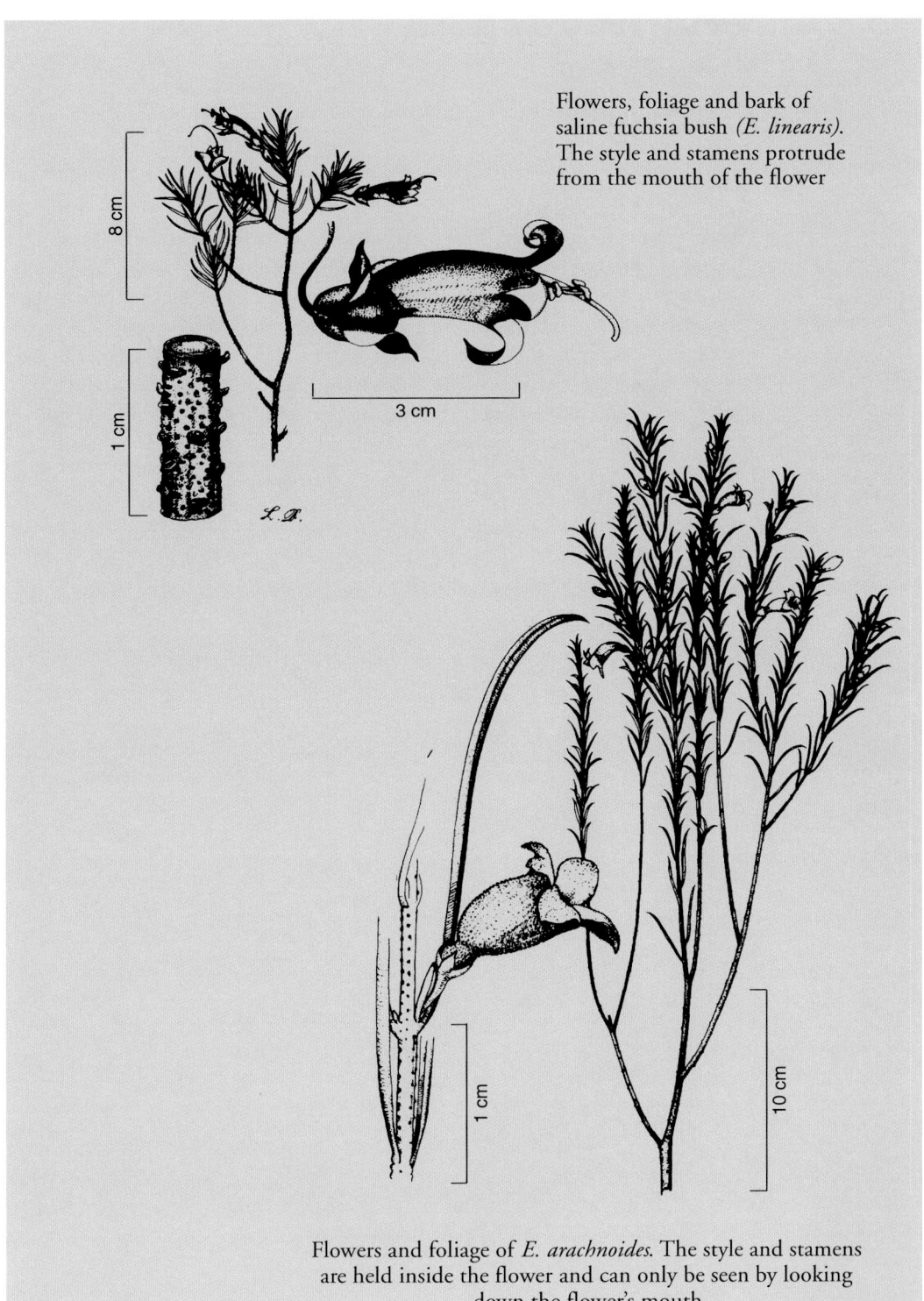

Flowers, foliage and bark of saline fuchsia bush *(E. linearis)*. The style and stamens protrude from the mouth of the flower

Flowers and foliage of *E. arachnoides*. The style and stamens are held inside the flower and can only be seen by looking down the flower's mouth

Flowers, foliage and bark of two eremophilas

POVERTY AND FUCHSIA BUSH GROUP

Felty fuchsia bush

Eremophila compacta

Family: Myoporaceae

GENERAL DESCRIPTION: Felty fuchsia bush is an upright shrub growing to 1 m. It is many-stemmed, with the lower parts of each stem retaining the bases of fallen leaves as rough scars. The leaves are a narrow, elliptical shape or may be broader at the tips and about 3 cm long and 7 mm wide. They are arranged densely in an alternate fashion along each stem and have a light greenish grey, felty covering, though the young growing tips are yellowish green. The flowers are borne on short stems at the ends of each of the branches. They are mauve, tubular in shape, about 2.5 cm long and with two upper and three lower lobes. The stamens and style are hidden within the flowers. The calyx supporting the flower is crimson in colour and heavily covered with hairs. The fruit is a hairless, small berry.

Felty fuchsia bush, about 1 m high

HABITAT: Felty fuchsia bush grows on rocky slopes and lower footslopes with square-fruited wattle, currant bush and low mulla mulla. It is also found in creeklines.

FORAGE VALUE: It is a valuable pasture plant, much sought after by stock.

INDICATOR VALUE: Its presence in a community indicates good range condition. As it is eagerly sought after by sheep, grazing has removed it from most communities which suit its requirements.

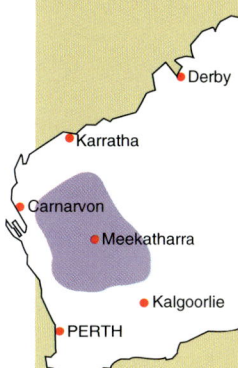

Tubular flowers, hairy, crimson sepals and felty leaves, about three-quarters actual size

POVERTY AND FUCHSIA BUSH GROUP

Waxy-leaf poverty bush

Eremophila 'crenulata' (manuscript name)

Family: Myoporaceae

GENERAL DESCRIPTION: Waxy-leaf poverty bush is a straggly shrub with a flat top, usually growing up to 1 m, but it can be higher. It is commonly found in dense stands. The shiny, dark green leaves are 2–4 cm long and up to 1 cm wide. They are oval to spear-shaped and have crinkled edges. The tubular flowers are up to 1.5 cm long and may be light blue or pink. The stamens and style extend slightly from the mouth of the flower. The petal tubes quickly fall off, leaving five pink or bluish sepals surrounding the developing fruit. The seeds are held in a hard berry, up to 1 cm in diameter.

HABITAT: Waxy-leaf poverty bush grows on alluvial plains with dark red or reddish brown sands over clays or on gradational soils. It grows with crinkled cassia, prickly acacia, currant bush and tall saltbush.

Waxy-leaf poverty bush, about 1.5 m high

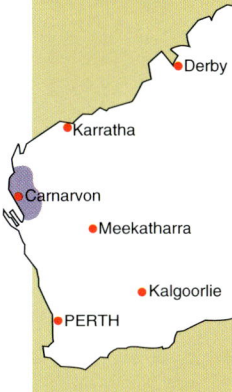

FORAGE VALUE: Waxy-leaf poverty bush has little forage value. It is never noticeably grazed.

INDICATOR VALUE: This species invades overgrazed Gascoyne bluebush pastures and in large stands indicates poor or very poor range condition. It is particularly noticeable on the inter-bank areas between sand banks on the Carnarvon coastal plain where it has replaced more valuable perennial shrub species.

Flowers, about actual size

POVERTY AND FUCHSIA BUSH GROUP

Royal poverty bush   Wedge-leaved eremophila, Pinyuru

Eremophila cuneifolia

Family: Myoporaceae

GENERAL DESCRIPTION: Royal poverty bush is a spreading shrub growing to 1.5 m, but more commonly reaching only 1 m. It has a resinous and rough appearance; the bark is dark grey and much roughened where the leaf bases have been shed. The olive-green, wedge-shaped leaves are up to 12 mm across, thickened, warty in appearance and have a covering of very short hairs. The tubular flowers are a deep purple or violet. They are borne very freely and bushes in flower are spectacular. After the flowers fall, the mauve-coloured sepals enlarge and persist, providing a prolonged colourful display. The fruit is small, rarely exceeding 5 mm.

Royal poverty bush, about 80 cm high

HABITAT: Royal poverty bush favours stony uplands with loamy soils that tend to be alkaline. It is tolerant of slightly saline conditions. It is found with low cassias and very sparse stunted mulga and occasionally with sago bush.

FORAGE VALUE: Royal poverty bush has no forage value. It does not appear to be eaten by stock.

INDICATOR VALUE: It has no indicator value.

Flower, about one and a half times actual size

POVERTY AND FUCHSIA BUSH GROUP

Frontage poverty bush

Eremophila delisseri

Family: Myoporaceae

GENERAL DESCRIPTION: Frontage poverty bush is a dense, green, rounded shrub reaching 0.4–1 m. Its stems are green and have a thick covering of short hairs. Many of the branches die at the tip and become spiny. Growing branches have their tips covered with hairs that produce a sticky secretion. Dust collects on the tips, often giving them a brown appearance. The leaves occur in opposite pairs, are spear-shaped and have a sparse covering of short, white hairs. They are about 1.5 cm long and 6 mm wide. The purple flowers are tube-like, up to 3 cm long and 8 mm wide. Each tube terminates in five equal lobes, which enclose the style and stamens. The flowers have long hairs on the inside and are covered on the outside by shorter hairs. The fertilised flower produces a hard, hairy fruit up to 6 mm long and 3 mm wide.

Frontage poverty bush, about 60 cm high

HABITAT: Frontage poverty bush occurs on alkaline loams and clays in frontage country and on the surrounds of lakes subjected to periodic flooding.

FORAGE VALUE: It is eaten by stock when the more attractive forages such as bindiis and saltbushes have been consumed.

INDICATOR VALUE: Its predominance indicates fair pasture condition. Under extreme grazing it will disappear.

Flowers, about actual size

POVERTY AND FUCHSIA BUSH GROUP

Wilcox bush Felt bush

Eremophila forrestii

Family: Myoporaceae

GENERAL DESCRIPTION: Wilcox bush is a compact, upright shrub growing to 2 m. It has light green to yellow growing tips and greyish green mature leaves. The leaves are oval to oblong, up to 1.5 cm long and 1 cm wide. They are covered in a dense mat of hairs that gives the plant a dull appearance. It usually flowers in winter and has a pale pink to light brown, tubular flower, up to 2 cm long with dark purplish spots down the throat. The stamens and style extend about 1 cm outside the flower. The seeds are borne in a fruit up to 5 mm in diameter.

HABITAT: Wilcox bush is most prominent growing in dense stands on the deeper sandy soils of wanderrie banks, on loamy sands receiving run-off water and on some sandplains in association with spinifex. It is usually associated with an under-storey of wanderrie grass and an over-storey of mulga or bowgada.

Wilcox bush, about 1.3 m high

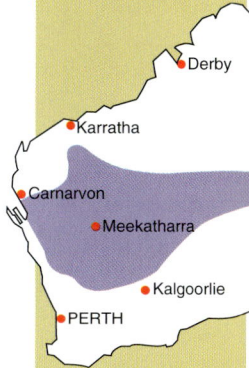

FORAGE VALUE: This shrub varies in its palatability to sheep from area to area, but it is rarely eaten by cattle. It is an important forage shrub in hard mulga and wanderrie country and on limestone country in the Carnarvon area. It is most readily eaten when it has fresh shoots and there is no ground feed. During dry conditions it retains its leaves, but is not eaten in these periods. The crude protein content varies from 9 to 17 per cent and the digestibility varies from 51 to 68 per cent.

INDICATOR VALUE: Its absence in wanderrie and hard mulga country indicates poor range condition. It is a reliable indication of good range condition in those environments, provided that the grazing has been by sheep. It is of little value as an indicator of range condition where the stock grazing are cattle.

Flowers, about actual size

POVERTY AND FUCHSIA BUSH GROUP

Turpentine bush Burra

Eremophila fraseri

Family: Myoporaceae

GENERAL DESCRIPTION: Turpentine bush is a bright green shrub which grows to 3 m. It has a very rough and scaly, greyish black bark. The leaves are up to 8 cm long, and up to 2 cm wide. They are bright green, very resinous and without hairs. The five-lobed, tubular flower is brownish red on the outside and pale lilac and finely spotted inside. Two of the four stamens hang outside the end of the flower. After the petal tube drops, the sepals supporting the fruit enlarge and provide a colourful show long after the flowers have dropped. They are pink to red, resin-covered and petal-like in form. The fruit is about 5 mm across and is a tapering egg-shaped structure on which the style persists for some time.

HABITAT: Turpentine bush is found on floodplains, in watercourses and on stony plains and hills. It occurs with mulga, hop mulga, tall sida, flannel bush and cassia. It appears to favour shallow hardpan soils rather than deeper sandy loams, and is a common plant on intergroves and alluvial plains.

Turpentine bush, 1.8 m high

FORAGE VALUE: It has little forage value though sheep, when hard pressed, will eat the plant sparingly. The flowers may also be eaten.

INDICATOR VALUE: Turpentine bush can take the place of more desirable species. When present to the exclusion of more desirable species, it is an indicator of poor range condition. Neither of the two closely related species have indicator value.

NOTE: Two species closely related to turpentine bush are found in the area. *Eremophila 'flaccida'* (manuscript name) is a 60 cm high, dense, spreading shrub growing on rocky hills in the northern Gascoyne, whilst *E. ramiflora* is a 2.5 m high, straggly, tufted shrub growing on hardpan plains in the eastern Murchison and north-east Goldfields.

Flower, about actual size

POVERTY AND FUCHSIA BUSH GROUP

Rock fuchsia bush   Stony poverty bush

Eremophila freelingii

Family: Myoporaceae

GENERAL DESCRIPTION: Rock fuchsia bush is an erect shrub with a height of 1–2 m. It tends to have a flattened crown above branches which are much roughened by the remains of old leaf bases. The leaves are a dull green, about 10 cm long, up to 6 mm wide and taper to a fine point. The young leaves are covered with dense down. Most leaves are pendulous and this accentuates the flattened appearance of the bush. The five-segmented, tubular flowers are lilac or pale blue and are usually borne singly on short and sticky stalks which arise in the leaf axils. The flowers are up to 3 cm long and the stamens protrude from the tube. The sepal lobes are covered with long, silky hairs and remain unchanged in size after the flower itself falls to the ground. The fruit is egg-shaped and up to 7 mm long.

Rock fuchsia bush, about 70 cm high

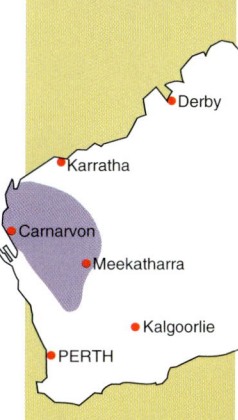

HABITAT: Rock fuchsia bush is found on rocky hillslopes and ridges and in other stony situations, where it grows with stunted mulga and low curara in these very arid environments. It is rarely, if ever, found on alluvial plains.

FORAGE VALUE: It does not appear to have any forage value.

INDICATOR VALUE: Rock fuchsia bush has no indicator value.

Flowers, about actual size

POVERTY AND FUCHSIA BUSH GROUP

Fine-toothed poverty bush

Eremophila georgei

Family: Myoporaceae

GENERAL DESCRIPTION: Fine-toothed poverty bush is an open shrub growing to about 1 m. The stems are a light brown, smooth and hairy. The leaves are a bright mid-green and have a sparse covering of white hairs. They are 1.5–2.5 cm long and 8 mm broad, oblong in shape, sometimes broader at the tip or pointed at the end. The margins of the leaves are finely toothed. The flowers are deep lilac to light purple in colour and tubular in shape with two upper lobes and three lower lobes at the end. They are borne on long, bent stems. The stamens and style are hidden within the flower. The sepals supporting the flower are tinged with purple. The seed case is hairy.

Fine-toothed poverty bush, about 1 m high

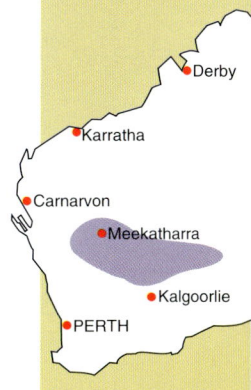

HABITAT: Fine-toothed poverty bush is common in mulga communities on lighter soils.

FORAGE VALUE: Fine-toothed poverty bush is rarely eaten by stock and therefore has no forage value.

INDICATOR VALUE: It has no indicator value.

NOTE: *Eremophila clarkei* is similar in leaf form to fine-toothed poverty bush, having the fine-toothed edge, but it can grow to 1.5 m. It grows in similar situations and also on rocky loams. The flowers are different in colour, however, and may be white, pale blue or pink, but have the same long, bent stem. Leaves, stems and calyx in *E. clarkei* are often resinous.

Flowers and leaves, about actual size. Note long, bent flower stalk and finely toothed margins to the leaves

POVERTY AND FUCHSIA BUSH GROUP

Tar bush

Eremophila glabra

Family: Myoporaceae

GENERAL DESCRIPTION: Tar bush is a shrub ranging from 0.5 to 2 m high. The foliage is extremely sticky to the touch. The light green leaves are 1.5–5 cm long and 5–8 mm wide. They are elliptical and are loosely scattered along the stem. The flowers are greenish yellow to red and tubular with five petal lobes, 2.5–3 cm long, and held on 2 mm long stalks. The upper four petal lobes are short and pointed, whilst the lower lobe forms a long lower lip that curls back from the flower. The stamens and style are held well outside the flower. The fruit is a berry and is up to 1 cm long.

HABITAT: Tar bush is found in sparse populations in mulga wash country and lake fringes and other habitats receiving run-on water. This species tolerates moderately saline soils.

Tar bush, about 1.5 m high

FORAGE VALUE: It is eaten readily by stock, especially by sheep during periods of forage shortage.

INDICATOR VALUE: It is a decreaser species and its presence indicates fair to good condition.

REHABILITATION: Tar bush may be used in rehabilitation programs if it occurs naturally in the area. Seed gathering may be tedious. An effective method is to shake the seeds off the bush onto a calico sheet spread on the ground.

Flowers and leaves, about three-quarters actual size

POVERTY AND FUCHSIA BUSH GROUP

Pink poverty bush

Eremophila 'glandulifera' (manuscript name)

Family: Myoporaceae

GENERAL DESCRIPTION: Pink poverty bush is a dense, grey shrub which grows to 1 m. The branches are rough as they retain the scars of the bases of leaves which fall. The leaves are covered with dense, fine hairs and are a bluish green, though the young growth can be yellowish. They are up to 10 mm long and up to 7 mm wide, elliptical in shape or broader towards the tip. Most leaves are held towards the ends of the stems. The flowers are produced along the stem in the leaf axils and are held on a very short flower stalk 1–4 mm long. Each flower consists of a red or pink tube, 8–20 mm long, which terminates in five lobes. The style and stamens protrude from the mouth of the flower. The fruit is a small, pale brown berry, 5–7 mm in diameter.

Pink poverty bush, about 70 cm high

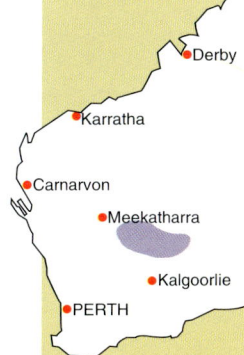

HABITAT: Pink poverty bush grows on mulga plains underlain by hardpan. It tends to grow beneath mulga and other tall shrubs.

FORAGE VALUE: It is generally not eaten by stock unless they are stressed.

INDICATOR VALUE: In heavier soils in mulga country it increases under intense grazing at the expense of cotton bush, but it is a normal component of sandy mulga communities.

NOTE: Pink poverty bush can be confused with Wilcox bush and felty fuchsia bush due to the similarity of the foliage. However, the pink flowers of pink poverty bush distinguish it from Wilcox bush, which has salmon-coloured flowers, and felty fuchsia bush, which has mauve flowers.

Rose pink flowers on short stems, and felty blue-green leaves, about actual size

POVERTY AND FUCHSIA BUSH GROUP

Thin-leaved poverty bush

Eremophila granitica

Family: Myoporaceae

GENERAL DESCRIPTION: Thin-leaved poverty bush is a shrub growing to 1.5 m. It is open in appearance and has delicate branches and stems. The bark is brownish and generally smooth. The leaves are bright green and quite narrow, being up to 4 cm long and 2 mm wide. They are sticky to the touch and are not stiff. The tubular flowers are mauve, white or pink and are borne on stalks up to 2 cm long. The stamens are carried entirely within the tube. The fruit is egg-shaped and lightly hairy. It does not exceed 5 mm in diameter.

HABITAT: It is most common in wanderrie communities where it forms part of a shrub layer below the mulga and above the grasses. It tends to favour sandy, acid soils and is not found in alkaline or loamy soils. It may be found on sandy outwashes below granite domes and tors.

Thin-leaved poverty bush, about 1 m high

FORAGE VALUE: It is not attractive to stock and has little forage value. The leaves have a strong aromatic odour which may make it unpalatable.

INDICATOR VALUE: Unlike most other poverty bushes, this species appears to be short-lived, but large numbers of seedlings appear after good seasons. With the onset of moisture stress it drops its leaves, and if the drought period is at all prolonged, it will die; it appears to have little drought resistance. This drought-evasive characteristic and its unpalatability confer little indicator value on the species.

Flowers, about actual size

POVERTY AND FUCHSIA BUSH GROUP

Murchison River poverty bush

Eremophila laanii

Family: Myoporaceae

GENERAL DESCRIPTION: Murchison River poverty bush is a dense shrub up to 3 m high. The leaves are a mid-green and are 3–4 cm long and 3–6 mm wide. The flowers are clustered towards the ends of the branches and are held singly on short stems, 3–6 mm long. They consist of a white or pink tube 2–3 cm long with five petal lobes. The style and stamens protrude beyond the lips. Each flower produces a berry-like fruit about 1 cm in diameter. It is flattened on one side and has a pointed apex.

HABITAT: Murchison River poverty bush is restricted to the floodplains of the Murchison River and may dominate in places.

Murchison River poverty bush, about 1.5 m high

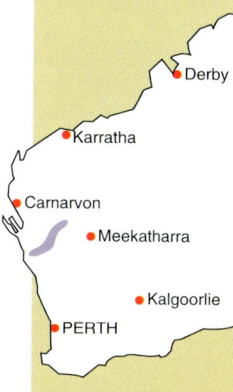

FORAGE VALUE: It is eaten by sheep and goats during times of forage shortage. Its palatability to cattle is unknown.

INDICATOR VALUE: It is a decreaser species and its presence in the shrub component of the pastures indicates fair to good condition.

Leaves and flowers, white form, about one-third actual size

Leaves and flowers, pink form, about one-third actual size

POVERTY AND FUCHSIA BUSH GROUP

Warty fuchsia bush   Warty-leaf eremophila

Eremophila latrobei

Family: Myoporaceae

GENERAL DISTRIBUTION: Warty fuchsia bush is a grey or green shrub which grows to 2 m. The bark is greenish brown and may be slightly scarred by the remains of leaf bases. The leaves are arranged alternately along the stem. They are finely hairy, linear to oblong, up to 4 cm long and 4 mm wide. They have margins which roll under and are rather paler underneath than on the upper surface. They are frequently warty in appearance. The tubular flowers range from yellow and orange to red. The flowers are held on short, hairy stalks in the axils of leaves and are up to 3.5 cm long, but rarely more than 1 cm across at the open end. The stamens are long and extend up to 2 cm beyond the mouth of the flower. The fruit is up to 1 cm long and pointed.

HABITAT: Warty fuchsia bush is found in mulga communities on sandy soils and loams as well as on rocky ridges and slopes.

Warty fuchsia bush, about 1.3 m high

FORAGE VALUE: Warty fuchsia bush is grazed by stock. Although it has been suspected of stock poisoning, this has never been confirmed. It is assumed, therefore, that it is not dangerous under field conditions. The plant tends to be rare in the grazed situation and returns to mulga communities only after they are protected from grazing. The few seeds produced and the infrequent germinating rains militate against a rapid return by the species.

INDICATOR VALUE: Vegetation which has warty fuchsia bush as a vigorous member of the shrub component is in good range condition. On the other hand, the absence of warty fuchsia bush need not indicate poor range condition, though the presence of stunted, overgrazed individuals indicates overuse.

NOTE: Slender poverty bush, or desert pride (*E. eriocalyx*), may be confused with warty fuchsia bush as it also has warty leaves. It can be distinguished by the hairy sepals which surround the flower and by the faintly hairy stems. Slender poverty bush is also a decreaser species and is eaten during periods of fodder shortage, particularly by goats.

Slender poverty bush, about actual size. Note persistent hairy calyx from which flowers have fallen

Flower of warty fuchsia bush, about actual size. Note the warts on the leaves

POVERTY AND FUCHSIA BUSH GROUP

Saline fuchsia bush   Harlequin fuchsia bush

Eremophila linearis

Family: Myoporaceae

GENERAL DESCRIPTION: Saline fuchsia bush is a compact dense leafy shrub which grows to 3 m. The bark is brown and slightly marked with the scars of old leaf bases. The leaves are bright green, sticky, hairless (except when young) and up to 5 cm long and 5 mm wide, and occasionally narrowing to a blunt point, but more commonly simply blunted. The tubular flowers are pink to red, yellow inside the throat, and are borne on stems up to 2 cm long. The sepals surrounding the flower are green and enlarge greatly when the flower falls. The seed case is egg-shaped, hard, wrinkled and much shorter than the sepals.

HABITAT: Saline fuchsia bush is a distinctive shrub on sluggish, saline drainages within undulating stony uplands and may also occur in crabholes within saline alluvial drainages. Its bright green colour and even shape contrast with

Saline fuchsia bush, about 1.7 m high

the lower and duller vegetation about it. As it occurs on saline inclusions within broader stony tracts and is unpalatable, it is frequently the dominant shrub in these situations, and is therefore very prominent.

FORAGE VALUE: Saline fuchsia bush is unpalatable and is rarely eaten by stock. It may be grazed sparingly in periods of extreme stress.

INDICATOR VALUE: Saline fuchsia bush is a normal component of the vegetation in upland saline habitats. When it is the only shrub present or when it is the dominant shrub, that community is in poor range condition.

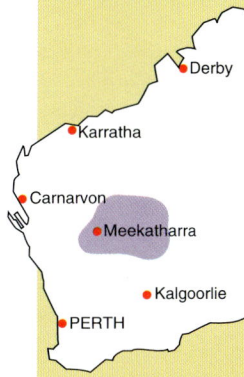

Flower and fruit, about actual size

POVERTY AND FUCHSIA BUSH GROUP

Grey turpentine bush

Eremophila macmillaniana

Family: Myoporaceae

GENERAL DESCRIPTION: Grey turpentine bush is an upright shrub growing to 2 m. The lower leaves are spoon-shaped while the upper ones tend to be spear-shaped. They are up to 5 cm long and 1.5 cm wide and have a covering of fine, felty hairs. The tubular flowers are pink to dull red on the outside and yellow on the inside. They are almost hairless. The style and stamens sit at the mouth of the flower.

Grey turpentine bush, about 1.4 m high

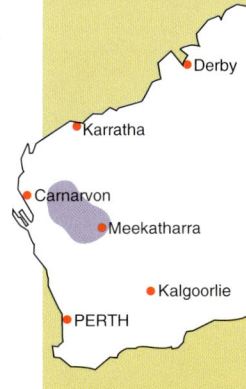

HABITAT: Grey turpentine bush grows on basalt and granite hills in shallow loam and clay soils. It grows with royal mulla mulla, cotton bush, warty fuchsia bush and mulga.

FORAGE VALUE: It is rarely eaten and, therefore, provides little feed for stock.

INDICATOR VALUE: It has no known indicator value.

Flowers, about actual size

POVERTY AND FUCHSIA BUSH GROUP

Fuchsia bush Travel bush, Emu bush

Eremophila maculata

Family: Myoporaceae

GENERAL DESCRIPTION: Fuchsia bush is a low, dense bush which generally grows to 50 cm, but occasionally reaches 1 m. It has dark green leaves up to 4 cm long and 6 mm wide. They have rounded or pointed ends and a glossy, hairless surface. The deep red to crimson tubular flowers are carried on 2 cm long stalks. The flower throats are often spotted with brown flecks. The style and stamens extend about 1 cm from the mouth of the flower. The seeds are borne in large, hard fruits, about 1 cm in diameter.

Fuchsia bush, about 80 cm high

HABITAT: Fuchsia bush grows on floodplains, valley floors, saline drainages, lake fringes and around crabholes in heavy loam or clay soils. It grows with sago bush, river saltbush, Gascoyne bluebush and other species tolerant of alkaline and often saline environments. It is rarely seen in non-saline environments.

FORAGE VALUE: It is poisonous if it is the sole diet of hungry stock.

It produces cyanide when eaten and is one of the most strongly cyanogenetic plants known. As little as 30 g of leaf will kill a sheep. However, stock appear to graze it with impunity in the normal grazing situation. It can contain up to 21 per cent crude protein and the digestibility is as high as 74 per cent.

INDICATOR VALUE: It is an indicator of fair range condition. Under heavy grazing more palatable species, such as river saltbush, may be grazed out, allowing this species to thicken up. Under very heavy grazing, fuchsia bush is also eliminated.

Almost mature fruits, about three-quarters actual size

Flowers, about three-quarters actual size

POVERTY AND FUCHSIA BUSH GROUP

Tall poverty bush

Eremophila maitlandii

Family: Myoporaceae

GENERAL DESCRIPTION: Tall poverty bush is an erect and many-branched shrub which grows to 3 m. The mature leaves are linear, up to 7 cm long and 5 mm wide, have blunt tips and a silver, felty covering. The growing tips are usually yellowish green. The tubular flowers are pale blue or lilac and up to 2 cm long. They are sparsely hairy on the outside and are borne on long stalks. The stamens and style remain inside the flower. The calyx beneath the petal tube has five blunt lobes and is distinctly felted in appearance. The seed is held in a fruit that is up to 5 mm in diameter.

Tall poverty bush, about 2 m high

HABITAT: Tall poverty bush grows on sandy banks and sandplains with deep, sandy soils and occurs with limestone wattle, bowgada, sago bush and WA buffel grass.

FORAGE VALUE: The palatability varies from area to area and may be soil related. Where it is eaten, it is grazed during periods of forage shortage. Its foliage persists well during droughts. Crude protein content varies from 9 to 16 per cent, and the digestibility varies from 50 to 62 per cent.

INDICATOR VALUE: Tall poverty bush is not a reliable indicator of range condition since it is so variable in palatability.

Flowers, about actual size

POVERTY AND FUCHSIA BUSH GROUP

Sandbank poverty bush

Eremophila margarethae

Family: Myoporaceae

GENERAL DESCRIPTION: Sandbank poverty bush is an open, often spreading shrub up to 1.3 m. The grey leaves which have a fine, felty covering are up to 5 cm long and 3 mm wide. When young they are a sulphur yellow colour. The mauve flowers are tubular and have a fine, hairy covering. The style and stamens are held just inside the mouth of the tube. The five sepals supporting each flower are covered in dense, woolly hair. The fruit is small.

HABITAT: Sandbank poverty bush grows on wanderrie sand banks with deep, acidic, sandy loams. It grows with wanderrie grasses, Wilcox bush and mulga.

FORAGE VALUE: It is rarely eaten and provides little forage for stock even though it is very drought resistant and retains its leaves in times of severe moisture stress.

Sandbank poverty bush, about 70 cm high

INDICATOR VALUE: It is an increaser species and dense stands indicate poor range condition. Under heavy grazing, populations of the more palatable species, such as Wilcox bush, are reduced and replaced by this unpalatable species. In the western Gascoyne its place as an increaser on sandy banks is taken by crinkled leaf cassia.

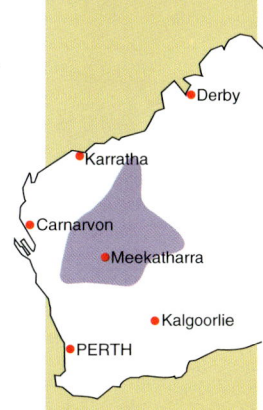

Flowers, about actual size

POVERTY AND FUCHSIA BUSH GROUP

Kopi poverty bush

Eremophila miniata

Family: Myoporaceae

GENERAL DESCRIPTION: Kopi poverty bush is a tall many-branched shrub which grows to 4 m. The stems are roughened from retaining the scars of the bases of leaves which have dropped. The leaves are narrowly elliptical and slightly sticky. They are mid-green in colour, 2–3 cm long and are carried at the ends of the branches and stems. The flowers are borne on long, curved stems and range from cream through yellowish green and scarlet to reddish brown in colour. They are often spotted on the inside of the tube. They are

Kopi poverty bush, about 1.5 m high

tubular and strongly divided into three pointed lobes. The stamens and style extend well beyond the mouth. The calyx is large and after the flower drops, it expands around the developing fruit.

HABITAT: Kopi poverty bush grows on alkaline soils around salt lakes, particularly on kopi dunes on the western sides of these water bodies where it is found with Cue york gum, dunna-dunna and limestone wattle.

FORAGE VALUE: It is rarely eaten and has no forage value.

INDICATOR VALUE: It has no indicator value for range condition assessment.

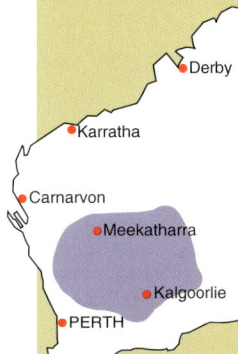

Leaves and flower, about actual size. Note residual scars of leaf bases and grouping of leaves at the ends of branches; the flower is distinctly lobed

POVERTY AND FUCHSIA BUSH GROUP

Granite poverty bush

Eremophila platycalyx

Family: Myoporaceae

GENERAL DESCRIPTION: Granite poverty bush is an erect shrub or small tree growing to 3 m. The leaves are grey or green, spear-shaped, up to 5 cm long, 6 mm wide and come to a sharp tip that forms a hook. The flowers are on 1.5 cm long stems, are tubular and either pink, orange or cream with black spots in the throat. The style and stamens sit in the mouth of the flower. Five mauve sepals persist after the petal tube falls and form a rosette behind the pointed fruit.

Granite poverty bush, about 1.3 m high

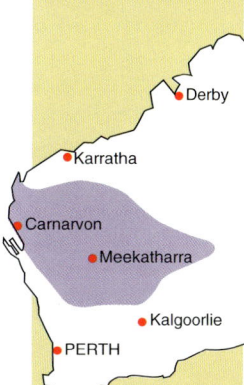

HABITAT: Granite poverty bush grows on granitic slopes and uplands with shallow loamy soils in association with mulga and cotton bush.

FORAGE VALUE: The plant contributes little to the diet of sheep or cattle, but is eaten by goats.

INDICATOR VALUE: It has no known indicator value.

Flowers and immature fruit, about one and a half times actual size

POVERTY AND FUCHSIA BUSH GROUP

Silver poverty bush   Wing-fruited eremophila

Eremophila pterocarpa

Family: Myoporaceae

GENERAL DESCRIPTION: Silver poverty bush is a dense, leafy shrub which grows to 3 m. The bark is dark brown and rough. The leaves are up to 3 cm long and 3 mm across, pointed and crowded together along the stems. They are a light grey and velvety to the touch. At a distance, silver poverty bush resembles a well-grown saltbush. The tubular flowers, which arise on moderately long stalks in the leaf and stem axils, are about 3 cm long and range in colour from pink to dull red on the outside and yellowish within. The outside of the flower is covered with dense, white, branched hairs. The sepals are small and do not enlarge. The fruit, which is prominently winged and covered with dense hairs, is usually about 1 cm in diameter and egg-shaped.

Silver poverty bush, about 2 m high

HABITAT: Silver poverty bush is a common shrub in saltbush and bluebush pastures on alluvial plains marginal to rivers and on drainages to salt lakes. The soils are usually alkaline and saline loams and clays with or without a hardpan.

FORAGE VALUE: Silver poverty bush is eaten by stock during periods of feed shortage. It is extremely drought tolerant and retains its leaves long into periods of severe moisture stress. It appears to be extremely long-lived.

INDICATOR VALUE: The presence of silver poverty bush indicates that a site has the capacity to support useful saltbush and bluebush species. When it is the only shrub species present, it can be assumed that the site is extremely degraded and in poor range condition.

Flower and fruit, about actual size

POVERTY AND FUCHSIA BUSH GROUP

Broom bush   Silvery emu bush

Eremophila scoparia

Family: Myoporaceae

GENERAL DESCRIPTION: Broom bush is an erect shrub growing to 2.5 m. It branches densely from the base. The leaves are opposite each other and are up to 2 cm long, 1 mm in diameter, round in cross-section and are coated with minute silvery scales. The tips come to a hooked point. The flowers are held on a very short stalk about 2 mm long. They are about 1 cm long and tubular, and are lilac in colour and have purple spots in the throat. The style and stamens are enclosed in the flower tube. The seeds are held in a small, pointed fruit, about 5 mm in diameter.

HABITAT: Broom bush grows in a wide variety of habitats on alkaline loams and clays beneath eucalypts in the Goldfields and on calcrete platforms adjacent to major watercourses elsewhere. In the former area it is associated with bladder saltbush and pearl bluebush and in the latter with mulga and pebble bush.

Broom bush, about 1.3 m high

FORAGE VALUE: It appears to be grazed in the Goldfields when young, but older plants are only slightly used by stock. It does not appear to be grazed in the Murchison. It is extremely drought resistant.

INDICATOR VALUE: Broom bush decreases under moderate to heavy grazing and an average stand of the species indicates fair condition in the Goldfields. Under extremely heavy grazing it becomes scarce, and its absence indicates poor condition. However, in overgrazed saltbush country, broom bush can become a woody weed. It can be eliminated by rabbits.

NOTE: Broom bush is similar to *E. pantonii* and *E. arachnoides* that are found in the Murchison, where these species are restricted to limestone.

Flowers, about actual size. Note that the leaves terminate in a hooked point

POVERTY AND FUCHSIA BUSH GROUP

Showy poverty bush

Eremophila spectabilis

Family: Myoporaceae

GENERAL DESCRIPTION: Showy poverty bush is a shrub which commonly grows to 1.5 m, but can be as high as 2 m. It has slender branches and dark grey bark. It has a variable covering of very short, stiff, white hairs which give it a slightly hoary appearance in some cases. The leaves have very short stalks (if any), are up to 8 cm long and 7 mm wide. The leaves are very aromatic, a common and distinctive trait in poverty bushes. The large flowers are borne on long stalks. They are pale to deep blue, tubular, up to 3.5 cm long and are frequently massed on the plant. The flower has five lobes at the end and one is rolled back, revealing the woolly throat of the tube. The stamens are entirely enclosed by the flower. The fruit is slightly pointed and covered with short hairs.

Showy poverty bush, about 80 cm high

HABITAT: Showy poverty bush is found in plant communities dominated by mulga. It favours deeper rather than shallow soils and is common in sandy soils and broad watercourses with soils at least 25 cm above the hardpan.

FORAGE VALUE: It is rarely eaten and has no known forage value.

INDICATOR VALUE: It has no known indicator value.

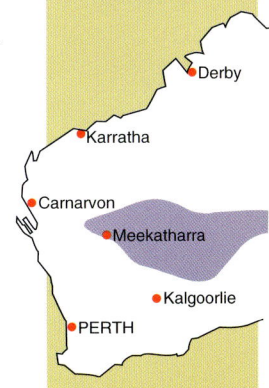

Flowers and a fruit, about actual size

Leafless ballart   Naked lady

Exocarpos aphyllus

Family: Santalaceae

GENERAL DESCRIPTION: Leafless ballart belongs to the same family as sandalwood. It is a shrub growing to 2 m. It is leafless except for the youngest branches, but the branches, even the oldest, are a bright green colour. The older stems have a scanty, thin, light grey bark and are somewhat ridged. Those leaves which are present resemble small scales, but they are hardly discernible. The very small flowers do not have showy petals and occur in short, dense spikes or clusters up to 4 cm long. The fruit is orange-red, egg-shaped and 3 mm long. It is borne on a fleshy stalk, which withers at maturity.

Leafless ballart, about 2 m high

HABITAT: Leafless ballart is found in saline habitats on sluggish drainages in stony uplands and in saltbush communities in saline valleys. It is at once distinguishable from other plants by its unusual appearance.

FORAGE VALUE: It does not appear to be eaten by stock.

INDICATOR VALUE: It has no indicator value.

Striped stems and small flowers, about actual size

Frankenia

Frankenia spp.

Family: Frankeniaceae

GENERAL DESCRIPTION: The frankenia group is very complex and it is difficult to give a satisfactory unified description. Most members are low, intricately branched, greenish grey shrubs up to 40 cm high and 60 cm wide. Some are small and herbaceous. The leaves are opposite each other on the stems, up to 3 mm long and 1 mm wide, and their edges are rolled downwards. The surface of the leaves and the stems are covered in salt crystals exuded by the plant. They flower in September/October and seed in October/November. The individual flowers consist of five petals up to 8 mm in diameter, which unite to form a tube up to 1 cm long in the lower two-thirds. The stamens are extended up to 2 mm above the petals. The flowers are generally held in dense heads and may be pink, cream or white. In some species the flowers tend to be less dense and in some are solitary. In the latter case, the plants themselves tend to be less woody.

Frankenia, about 20 cm high

HABITAT: Frankenias are found in saline areas and grow with a wide range of salt-tolerant shrubs such as golden bluebush, swamp saltbush, and samphire in open shrublands.

FORAGE VALUE: Frankenias contain up to 10 per cent salt and are only eaten if stock water contains little salt. Their foliage persists during droughts and they are a useful forage. The crude protein content ranges from 8 to 10 per cent and the digestibility ranges from 30 to 37 per cent.

INDICATOR VALUE: Frankenias are moderately resistant to grazing and are an indicator of good to fair range condition where the stock water contains less than 2000 ppm (140 grains per gallon) of salt. If the stock water has a higher salt content, frankenias have no indicator value.

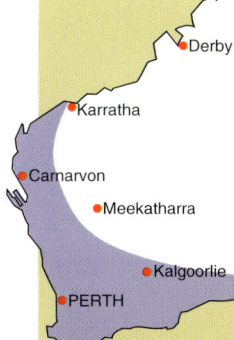

Flowers, about one and a half times actual size. Note the salt crystals on the foliage

Breelya   Kite leaf

Gastrolobium laytonii

Family: Papilionaceae

GENERAL DESCRIPTION: Breelya is a compact shrub, commonly reaching 2–3 m, but it can be up to 5 m high. It has multiple branches divided from the base. The leaves are greenish grey and opposite each other on the stem, with adjacent pairs alternating in direction. They are up to 5 cm long and 2 cm wide. The tip may come to a sharp point and there may be a pointed lobe on each side of the leaf. These give the leaves the characteristic kite leaf shape. The flowers are yellow and brown, 5 mm long and pea-flowered in shape. The shrub flowers from August to October. The seeds are about 1 mm in diameter and carried in 1 cm long, pointed pods.

HABITAT: Breelya grows mainly in thickets around granite domes and on the edges of breakaways. It is also found on gravelly rises within sandplains in the Yalgoo/Paynes Find area. It grows in association with miniritchie and *Dodonaea* spp.

Breelya, about 2 m high

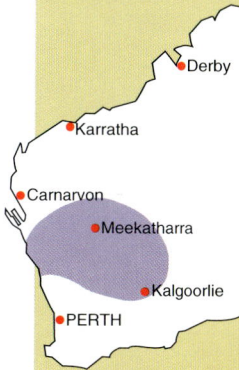

FORAGE VALUE: Breelya has no forage value and is very poisonous to stock as it contains a toxic substance similar to monosodium fluoroacetate (1080). It is generally eaten when it has fresh green shoots and when there is no other attractive forage. It has been responsible for large losses of sheep, cattle, goats and horses. The flowers and seeds are the most dangerous parts of the plant.

INDICATOR VALUE: It has no known indicator value.

Leaves of related species G. brevipes, *about actual size. Note absence of kite leaf characteristics;* G. brevipes *is probably also toxic*

Kite-shaped leaves, about half actual size

Red grevillea

Grevillea deflexa

Family: Proteaceae

GENERAL DESCRIPTION: Red grevillea is a dense, mid-green shrub up to 1 m high with needle-like leaves. The leaves are 3–4 cm long, up to 4 mm wide and sharply inrolled at the edges. They terminate in a 3 mm long, sharp point. The leaves are dark green on top and light green underneath. The flowers are generally red, but may also be yellow or orange in colour. They are held in spikes which hang down through the foliage. A smooth, red style protrudes from each flower.

Red grevillea, about 70 cm high

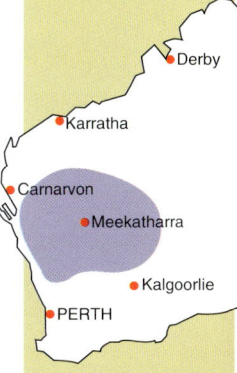

HABITAT: Red grevillea is always found in small isolated populations in and around granite outcrops and in mulga country.

FORAGE VALUE: It is eaten by stock in times of forage shortage.

INDICATOR VALUE: Red grevillea has no indicator value.

Flowers and two-toned leaves, about actual size

Cue grevillea

Grevillea inconspicua

Family: Proteaceae

GENERAL DESCRIPTION: Cue grevillea is a many-branched but fairly open shrub which reaches 2 m. Some plants may be 2 m across. The branches may be strongly pointed. The leaves are a bright green, undivided and linear in shape. They hang downwards from the stem and are up to 4 cm long. The unobtrusive flowers are pale pink and are borne in the axils between stem and leaf. The hollow, woody fruit is rounded and has a long, narrow beak. It is about 1 cm long and 5 mm wide. It opens when dry to release two winged seeds.

Cue grevillea, about 1 m high

HABITAT: Cue grevillea grows with mulga, turpentine bush, currant bush, cotton bush and flannel bush on soils derived from basic rocks such as basalts. It may also grow on footslopes below outcropping basic rocks, particularly in drainage lines associated with them.

FORAGE VALUE: It is a preferred species and is heavily grazed by sheep.

INDICATOR VALUE: Cue grevillea is a good indicator of range condition.

NOTE: Cue grevillea is listed as a rare and endangered species by the Department of Conservation and Land Management.

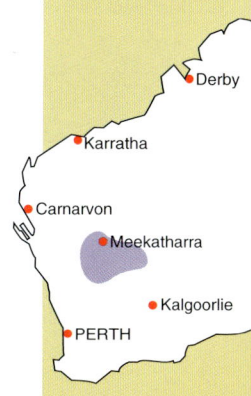

Leaves, pale flowers and fruit, about actual size

Sweet samphire   Sturt's pigface

Gunniopsis quadrifida

Family: Aizoaceae

GENERAL DESCRIPTION: Sweet samphire is a greenish, compact shrub up to 50 cm high and 80 cm in diameter. The branches are prominently forked in pairs. The leaves, which occur in opposite pairs, are 2–4 cm long, succulent and round in cross-section. The flowers have four white, triangular petals arranged in the shape of a cross. There are numerous stamens. The fruit pods are roundish, being 5–7 mm in diameter, and contain many small, black seeds.

HABITAT: Sweet samphire is found around saline lakes and river fringes, on medium to heavy textured soils. It grows with samphires and saltbushes and stands out from them with its bright green leaves and low, compact and well-leaved form.

Sweet samphire, about 40 cm in diameter

FORAGE VALUE: Sweet samphire is high in salt and varies locally in acceptability depending on the salinity of local stock water. Where the stock water is high in salt, it contributes little to diet since stock cannot tolerate the total salt load in water and feed. Where stock water is fresh or only slightly brackish, sweet samphire is a valuable component of the diet of sheep.

INDICATOR VALUE: Since its use is affected by the quality of the stock water, it can be an ambiguous indicator of range condition. Where the water is not saline, it is a useful measure of range condition, but with saline water it is not.

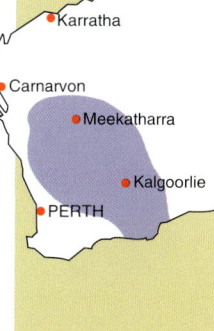

Flowers, one and a half times actual size

Samphire

Halosarcia doleiformis

Family: Chenopodiaceae

GENERAL DESCRIPTION: There are many different plants which go under the collective name of samphire. The plant illustrated here is one of 37 generally similar succulent species found in saline and flooded situations in Australia. This samphire, in common with others in the group, is characterised by its upright, succulent and leafless branches. Some grow to about 1 m, but others are more spreading in their habit and rarely exceed 50 cm. The stems are bright green or yellow, but may be a darker green or often greyish or even a dull red or brown. The flowers are embedded in the joints of the fleshy branches and are rarely seen. The stamens are visible when they protrude for a short time. The seed cases tend to be spongy and contain a very small seed, which falls when the 'fruiting stem' breaks up.

Samphire, about 50 cm in diameter

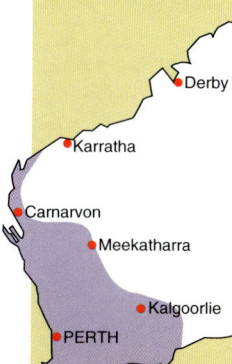

HABITAT: Samphires grow in saline situations which are sometimes flooded. They form the first fringing community above the salt line in salt lakes and are found in the flooded and saline areas of many drainages. They can also be found in the sluggish, saline drainages of the stony uplands, on the sea coast and in estuaries.

FORAGE VALUE: Samphires can contain up to 24 per cent salt and are only eaten if stock have access to fresh water. As this is uncommon in the communities in which it grows, samphire is rarely valuable as forage. Where the high salt levels can be tolerated the plants contribute usefully to the diet of grazing animals.

INDICATOR VALUE: Samphire can be used as an indicator of good range condition when non-saline water is available to stock, but it is of no indicator value when only salty water is available.

Succulent, leafless stems with decaying female flowers at the tips, about actual size

Dunna-dunna

Lawrencia helmsii

Family: Malvaceae

GENERAL DESCRIPTION: Dunna-dunna is a curious plant with a growth form resembling a small cactus. It grows to about 1 m and is many-branched. The leaves, which are found in tightly pressed bundles, completely surround the woody stem. They are less than 5 mm long. There are separate male and female plants. The flowers of each sex arise within the centres of the leaf bundles. When in flower, the yellow male flowers confer a golden appearance on the plant.

Dunna-dunna, about 80 cm high

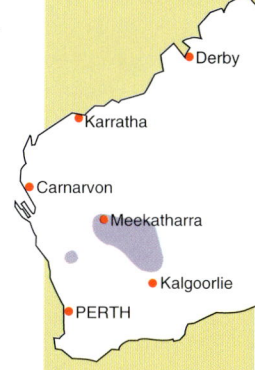

HABITAT: Dunna-dunna grows on the kopi or gypsum-based dunes and platforms on the margins of salt lakes.

FORAGE VALUE: It is extremely drought tolerant but is not eaten by stock.

INDICATOR VALUE: Dunna-dunna has no indicator value.

Portion of a dunna-dunna stem, showing tight bundles of leaves and female flowers, about one and a half times actual size

Grey fan leaf

Lawrencia squamata

Family: Malvaceae

GENERAL DESCRIPTION: Grey fan leaf is a low, intricately branched shrub which grows to 1 m, but is more commonly 30–60 cm high. The intertwined branches are stout and woody and are covered, as are the leaves, with tightly attached small scales. Some branches may be devoid of leaves and end in a tip. Others may be short, but are densely covered with leaves. The leaves are 3–8 mm long and are broader at the tip, which may be notched. They are green to grey in colour. There are separate male and female flowers and both are small. They are mostly clustered together on short shoots. The fruits consist of several small fruitlets.

Grey fan leaf, about 80 cm high

HABITAT: Grey fan leaf grows with saltbush, sago bush and sage on alkaline loams and clays which may also be saline. It is common around salt lakes.

FORAGE VALUE: It is eaten by stock.

INDICATOR VALUE: Grey fan leaf appears to replace plants such as saltbush and bluebush in salt lake and saline communities. It may be used as an indicator of past overuse when it is found as the dominant species in these communities.

Leaves and female flowers, some branches with no leaves, about half actual size

Water bush   Australian boxthorn

Lycium australe

Family: Solanaceae

GENERAL DESCRIPTION: Water bush is a dense, intricately branched, globular-shaped shrub growing to 2.5 m. The main branches and side branches often end in a spine. The leaves are variable, reflecting the seasonal conditions and are largest and most succulent after rain. They can be 3–25 mm long and 2–5 mm wide. They are oval in cross-section and are borne in groups on the stems. The flowers are small and tubular and are up to 12 mm long, with four white to pale mauve petals. Each flower produces a red, succulent, oval berry. Water bush can be confused at a distance with currant bush, which grows in similar fashion. Water bush is distinguished by its tubular rather than fan-shaped flower and the pronounced succulence of its leaves.

Water bush, about 1.2 m high

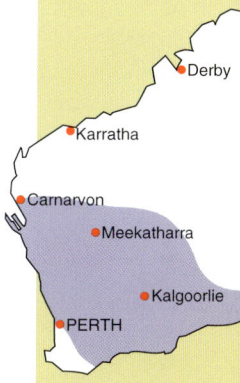

HABITAT: Water bush grows in association with sago bush on limestone areas with alkaline clay soils. It is not found on the more saline areas of lakes and valley floors, though it can occur on sand dunes which surround lake pans.

FORAGE VALUE: It is palatable but loses its leaves during dry times. The intricately laced branches prevent sheep eating all the available leaf material; this has probably favoured the survival of this species and its ability to colonise niches vacated by more palatable species in overgrazed pastures.

INDICATOR VALUE: Water bush is not a sensitive indicator of range condition. It is not a common plant in lake country and changes in the population do not reliably reflect changes in range condition.

Flowers, succulent foliage and spiny branches, about actual size

Red berries and succulent foliage, about actual size

Bluebush group

The bluebushes belong to the Chenopodiaceae family, which is noted for its salt-loving plants including saltbushes, samphires, and bindiis. Most bluebushes belong to the genus *Maireana* and are found in saline environments. However, some, such as mulga bluebush, are restricted to mulga shrublands which are not saline.

The bluebushes are generally low shrubs up to 1.5 m high and have small, fleshy leaves which are, in most cases, round in cross-section.

The most distinctive characteristic of bluebushes is the seed case. It is a woody structure not more than 5 mm across and surrounded by a papery wing which can be up to 2 cm in diameter. In some species small appendages or processes are attached to the seed case.

There are about 40 species of bluebush in Western Australia and most of these are found in the arid shrublands. Nineteen of the more common species are described in this book.

Illustrations from Wilson, P.G., 1975, 'A taxonomic revision of the genus *Maireana* (Chenopodiaceae)', *Nuytsia* 2, 1–83.

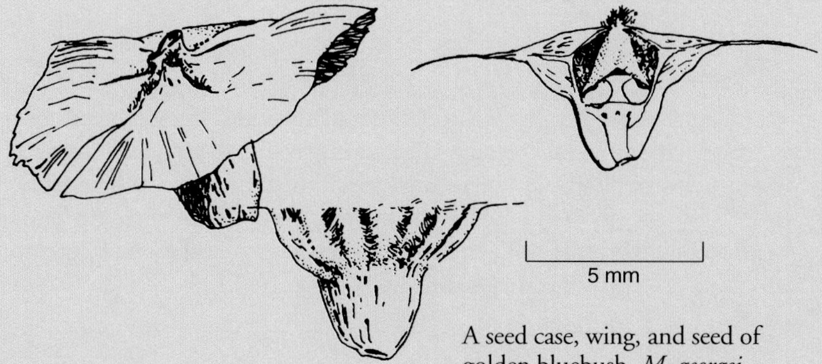

A seed case, wing, and seed of golden bluebush, *M. georgei*

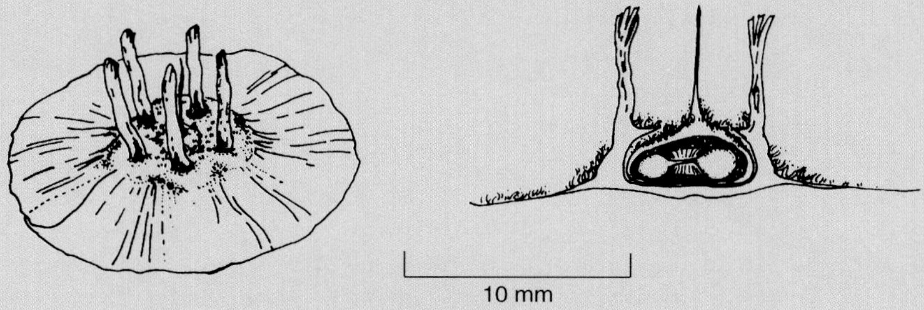

A seed case, wing processes and seed of ball-leaf bluebush, *M. glomerifolia*

A seed case, wings and seed of Gascoyne bluebush, *M. polypterygia*

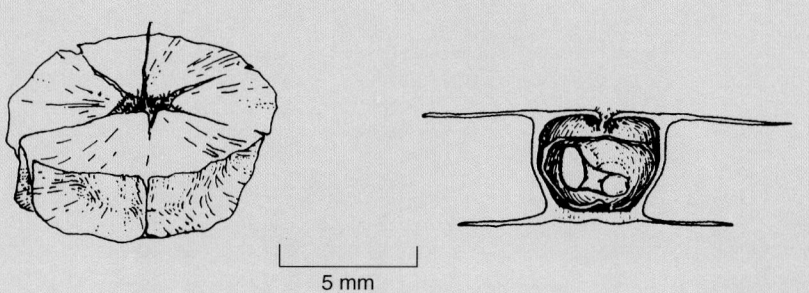

Seed cases of some bluebushes

BLUEBUSH GROUP

Brittle bluebush

Maireana amoena

Family: Chenopodiaceae

GENERAL DESCRIPTION: Brittle bluebush is a low, herbaceous or brittle shrub which can be 60 cm high. The young branches are obviously woolly. The leaves are fleshy and often hirsute and may be globular or slightly rounded. They are up to 1.5 cm long. The flowers are inconspicuous. The seed case is surrounded by five horizontal wings which are pink and frequently hairy. Each is up to 3 mm across.

Brittle bluebush, about 50 cm high

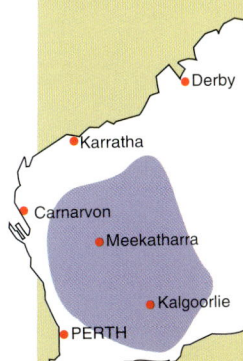

HABITAT: Brittle bluebush grows on the sandy rises surrounding salt lakes, saline drainages and saline plains marginal to floodplains. It readily re-colonises overgrazed country which is spelled.

FORAGE VALUE: It is readily sought after by sheep, but is not highly resistant to grazing.

INDICATOR VALUE: Strong stands of brittle winged bluebush indicate good range condition.

Distinctly hairy leaves, winged pink seed cases, about actual size

BLUEBUSH GROUP

Breakaway bluebush

Maireana aff. *tomentosa*

Family: Chenopodiaceae

GENERAL DESCRIPTION: Breakaway bluebush is a small shrub, usually less than 40 cm high and 40 cm wide. The whole plant has a covering of dense hairs. The leaves are 5–15 mm long and 1 mm wide and slightly succulent. The flowers are held in a spiral on the growing tips. The flowers are insignificant, but mature as fragile seed cases 2 mm in diameter containing one seed. A membranous wing or saucer up to 1 cm in diameter surrounds the seed case. It turns yellow on maturity.

Breakaway bluebush, about 30 cm high

HABITAT: Breakaway bluebush is usually found beneath breakaways on shallow, stony, clay soils.

FORAGE VALUE: It is a highly nutritious forage and is eagerly sought after by stock.

INDICATOR VALUE: It is a short-lived perennial. Its presence indicates good to fair condition.

Leaves and seeds, about actual size

BLUEBUSH GROUP

Spiny bluebush

Maireana aphylla

Family: Chenopodiaceae

GENERAL DESCRIPTION: Spiny bluebush is a compact shrub which grows to 1.5 m and up to 1.5 m wide. There may be several plants intertwined in a group. It has very small leaves which are 1–4 mm long and 1 mm in diameter. They are round in cross-section and fleshy. The short side branches lose their leaves early and remain bare, giving the plant its spiny appearance. The flowers are inconspicuous and occur in winter and spring. The seed cases which follow are surrounded by a flat, brown, papery wing which is up to 8 mm in diameter. The seed itself is held in the centre of the wing in a small, slightly woody structure.

Spiny bluebush, about 1.3 m high

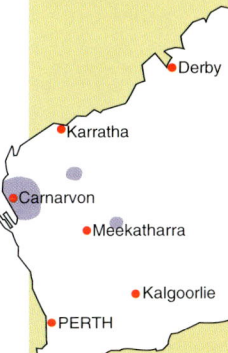

HABITAT: Spiny bluebush grows on alluvial plains, subject to inundation on soils which are generally cracking clays. It grows with fuchsia bush, Gascoyne bluebush and swamp saltbush.

FORAGE VALUE: Spiny bluebush is only eaten when animals are forced onto it by feed shortages. It is, however, resistant to grazing and maintains what little foliage it has, well into drought.

INDICATOR VALUE: It has little indicator value.

Winged seeds, succulent leaves and spiny side branches, about half actual size

BLUEBUSH GROUP

Five-pin bluebush   Bronze bluebush

Maireana atkinsiana

Family: Chenopodiaceae

GENERAL DESCRIPTION: Five-pin bluebush is a low shrub normally growing to 60 cm, but ungrazed plants may be taller. It has multiple, intricately interwoven branches. Spreading branches frequently root where they strike the ground. The bark is light grey or creamy grey. The leaves are grey-blue in colour, fleshy and have a sparse, hairy covering. They are usually 5–10 mm long, have a rounded end and are arranged alternately on the stem. There are separate male and female plants. The small flowers are produced in pairs on the female plants and are succeeded by prominent, pink bluebush seed cases bearing five lobes. Two pairs of oblong processes about 5 mm high are found on two of the five lobes and one on another. They are a key characteristic for this plant. Mature fruits change colour to bronze. The male flowers are inconspicuous.

Five-pin bluebush, about 60 cm high

HABITAT: Five-pin bluebush grows on the upper margins of saltbush communities with silver saltbush and narrow-leaved mulla mulla. It does not favour highly saline situations.

FORAGE VALUE: It is eagerly sought after by stock and tends to be rare in heavily grazed communities.

INDICATOR VALUE: Five-pin bluebush is an indicator of good range condition when it is common. Its absence from a community does not necessarily mean that range condition is poor as it is somewhat restricted in its distribution.

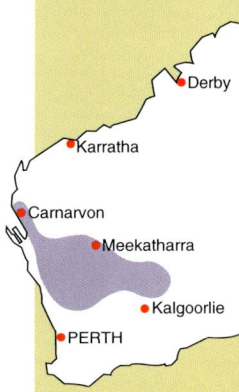

Fruits showing five processes on the wings of each, about actual size. Fruit turns bronze as it matures

Mature fruit, now faded to bronze, about actual size

BLUEBUSH GROUP

Short-leaf bluebush

Maireana brevifolia

Family: Chenopodiaceae

GENERAL DESCRIPTION: Short-leaf bluebush is an erect shrub which reaches about 1 m. The branchlets are thin and irregularly covered with fine hairs, which are often rolled into balls. The leaves are green, tubular in cross-section, fleshy and 2–5 mm long. For the most part they are hairless. The seed case is surrounded by five horizontal wings about 2–3 mm long. Each wing contains fine, brown veins. The fruits are borne very freely at the ends of the branchlets.

HABITAT: Short-leaf bluebush is not common north of the 26°S latitude. It is found in alkaline and saline soils, particularly around lake systems and on sluggish drainage lines. It is an early coloniser of disturbed situations and establishes rapidly in abandoned mining tenements.

Short-leaf bluebush, about 1 m high

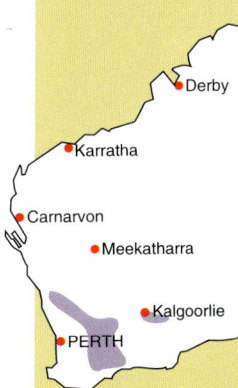

FORAGE VALUE: It is not readily eaten, but is used after annual species have been consumed. It contains up to 16 per cent crude protein and is low in fibre. The salt content may be as high as 15 per cent which would limit intake. It has performed well in grazing trials in the agricultural areas, provided that it is not grazed continuously.

INDICATOR VALUE: It has no indicator value.

REHABILITATION: It is a useful species for inclusion in primary rehabilitation mixes applied to resculptured dump faces provided that it is used where winter rainfall is dominant. It demands fairly reliable winter rain for successful establishment and growth. Seed should be gathered locally, as far as possible, and stored in air-dry conditions. A seeding rate of about 1 kg/ha of short-leaf bluebush is recommended for the mix.

Leaves and fruits, about three-quarters actual size

BLUEBUSH GROUP

Mulga bluebush

Maireana convexa

Family: Chenopodiaceae

GENERAL DESCRIPTION: Mulga bluebush is an open, graceful shrub which grows to 2 m. It has robust, woody lower stems from which rise the new and flexible younger growth. The branchlets of the new growth are covered with fine hairs. The leaves are alternate, round in cross-section and pointed towards the tips. They are up to 2 cm long, 1–2 mm thick and covered with silky hairs. The small flowers are borne in the axil between leaf and stem. They are succeeded by prominent, winged seed cases about 1.5 cm in diameter which are suffused with gold or bronze colours when mature. The fruit within the centre of the wing is hard and woody with a close, woolly covering on the top.

Mulga bluebush, about 60 cm high

HABITAT: Mulga bluebush grows in non-saline mulga communities with hardpan beneath sands and loams. It may also occur in saline communities, but it is not common.

FORAGE VALUE: Mulga bluebush is a valuable forage plant. It is fairly drought tolerant and loses its leaves only with severe moisture stress. It is palatable and provides a source of protein when annual feed has dried. It responds vigorously to small amounts of rainfall and, if rested after rain, produces a great deal of dry matter. Continual heavy grazing does not allow it to reach its potential. Experience in the Murchison has shown that it responds very favourably to annual or biennial rests of six months. This grazing practice allows the plants to increase in numbers as well as in size.

INDICATOR VALUE: Mulga bluebush is an important indicator of range condition.

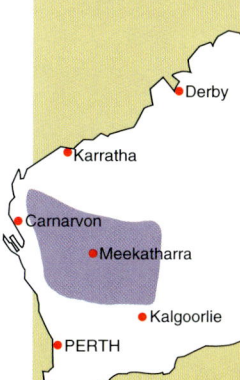

Large winged seeds and succulent leaves, about three-quarters actual size. Note white hairs on stems and leaves

BLUEBUSH GROUP

Golden bluebush George's bluebush

Maireana georgei

Family: Chenopodiaceae

GENERAL DESCRIPTION: Golden bluebush is a compact, rounded, stout-stemmed shrub that grows to about 50 cm. The branches are woolly and bear thin, succulent leaves which are 1–1.5 cm long. They are arranged alternately along the stems and are sparsely or densely woolly, but they can also be hairless. The flowers are small, bisexual and are found in the axils of stem and leaf. The seed case which succeeds the flower is very prominent. The papery wing is golden, up to 2 cm across, and surrounds the woody seed capsule, which is 4 mm in diameter.

HABITAT: Golden bluebush is found in the saline stony uplands, basalt uplands, beneath breakaways on saline alkaline soils and also occurs in saline shrub communities.

Golden bluebush, about 40 cm across

FORAGE VALUE: Golden bluebush is a valuable fodder shrub. It contains 13–26 per cent crude protein and has a digestibility of 58–67 per cent.

INDICATOR VALUE: It is a decreaser species and is a reliable indicator of good range condition. Studies in the Murchison have shown that it sets seeds only after winter rain and that these seeds are viable for a year. Re-establishment will only occur, therefore, in the year immediately following good winter rains. Seed production, however, is very dependent upon the amount of grazing received. Plants heavily grazed before winter rains tend to produce very little seed. If satisfactory regeneration of golden bluebush is to occur, pastures containing the parent plants need to be rested in the summer before the winter rains, so that sufficient seed will be produced.

REHABILITATION: Golden bluebush is a useful species for rehabilitation work. Seed may be collected locally in large quantities. It must be dried thoroughly and used within a year, as the seeds are prone to fungal attack.

Large, gold-coloured winged seeds and succulent leaves, about actual size

BLUEBUSH GROUP

Ball-leaf bluebush

Maireana glomerifolia

Family: Chenopodiaceae

GENERAL DESCRIPTION: Ball-leaf bluebush is a low, rigidly branched shrub which grows to 40 cm. The bluish green leaves are very small and are tightly clustered in small balls along the stems and branches. From a distance the plant appears to consist of leafless, green stems. The flowers are also very small and inconspicuous, and occur on very short spikes. They are followed by small reddish seed cases, which are closely attached to the branches. In common with most bluebushes, the seed cases are surrounded by a horizontal membranous wing, in this case 1.5 cm across. There are a number of delicate oblong processes standing above the wing.

Ball-leaf bluebush, about 50 cm across

HABITAT: Ball-leaf bluebush occurs on floodplains, salt lake fringes, basalt slopes, stony upland plains and particularly on slopes beneath breakaways. It grows with river saltbush, sago bush, golden bluebush and bladder saltbush.

FORAGE VALUE: It is moderately palatable to stock, but since it tends to occupy specialised sites which are attractive to stock, it is subject to very heavy use. It is rare, therefore, to find communities of ball-leaf bluebush in good condition. Studies have shown that it is also a species favoured by kangaroos, which, when present in sufficient numbers, will prevent its re-establishment.

INDICATOR VALUE: It is very resistant to grazing, but it decreases under heavy grazing. Its absence indicates poor range condition.

Pink winged seeds and leaf balls, about half actual size

BLUEBUSH GROUP

Woolly bluebush

Maireana lanosa

Family: Chenopodiaceae

GENERAL DESCRIPTION: Woolly bluebush, a weak perennial shrub with a woody rootstock, grows to 50 cm. It has straggly branches covered with woolly hairs. The leaves are narrow, up to 12 mm long, flattened, semi-succulent and covered with woolly, oblique hairs. The flowers are borne in winter and spring and are hidden between the stems and leaves. Each gives way to a woolly, leathery seed case, which is surrounded by a flat, greenish white wing about 8 mm in diameter. The seed case has six thin vertical processes above the wing and their prominence distinguishes it from most other bluebushes.

Woolly bluebush, about 30 cm high

HABITAT: Woolly bluebush is found on sandy banks and plains with limestone wattle, Wilcox bush and grey cassia.

FORAGE VALUE: It is not often eaten by stock.

INDICATOR VALUE: It has no indicator value.

NOTE: There is a second form of woolly bluebush, found growing on saline plains and on the margins of streamlines in alkaline and salty soils. It is very woody, gnarled and rarely exceeds 50 cm. The branches are stiff and erect. This variant is extremely attractive to stock and is commonly removed by overuse. Its presence indicates good range condition.

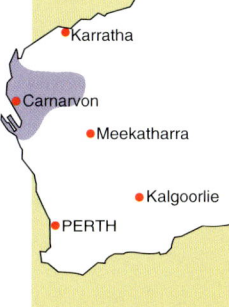

Winged seed case with vertical processes covered with woolly hairs, about one and a half times actual size

BLUEBUSH GROUP

Pussy bluebush

Maireana melanocoma

Family: Chenopodiaceae

GENERAL DESCRIPTION: Pussy bluebush is a hairless shrub up to 30 cm high with large, black, whiskery seed cases. The branches are slender and marked with long, parallel lines. The leaves are up to 1 cm long, thin, round in cross-section, succulent and have a grey, waxy covering. The flowers are small and insignificant, but develop into seed cases that are up to 5 mm wide and surrounded by a circular, thin, papery wing which can be 1.5 cm in diameter. Numerous bristles, up to 7 mm long, protrude from the top of the seed case. They are yellow when immature and turn black on maturity.

HABITAT: Pussy bluebush grows on hills and rocky uplands which have thin, skeletal soils underlain by rock or hardpan. It grows with mulga and many species of poverty bush.

Pussy bluebush, about 20 cm high

FORAGE VALUE: It is very palatable to stock.

INDICATOR VALUE: It is a decreaser species and rapidly disappears even under moderate grazing. It can be used as an indicator of range condition on stony upland pastures.

NOTE: Pussy bluebush has the only bluebush seed case with numerous bristles and can be readily distinguished by this characteristic.

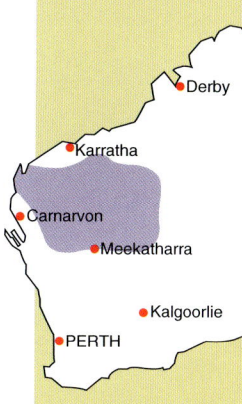

Slightly immature seeds, about actual size

Mature seeds, about actual size

BLUEBUSH GROUP

Flat-leaved bluebush   Low bluebush

Maireana planifolia

Family: Chenopodiaceae

GENERAL DESCRIPTION: Flat-leaved bluebush is a grey, open shrub which grows to 1 m, but is commonly smaller. Unlike most other bluebushes, it has flattened and rather thin leaves, up to 1 cm long and about 4 mm wide, and arranged alternately along the stem. The leaves are generally succulent and are usually covered with a mat of fine, grey hairs. Some types, however, are only sparsely hairy. The flowers are very small and are borne at the ends of the branches. The seed case is not robust, tending to be thin-walled. It is surrounded by a membranous wing up to 1.5 cm in diameter. It turns a pale straw colour when mature.

Flat-leaved bluebush, about 40 cm high

HABITAT: Flat-leaved bluebush grows in mulga shrublands on acid loams with mulga, Wilcox bush and cotton bush. It avoids the more saline and alkaline environments favoured by other bluebushes and saltbushes. It can also occur on the rocky slopes of hills and low ranges and may be found on the margins of some saline valley floors.

FORAGE VALUE: It is very palatable and is preferentially grazed by stock, particularly at the onset of dry periods when annual species have disappeared. It is very drought resistant and, therefore, an important browse plant for stock.

INDICATOR VALUE: Populations increase under moderate grazing pressure, but decline under heavy grazing pressure. It is an indicator of fair range condition.

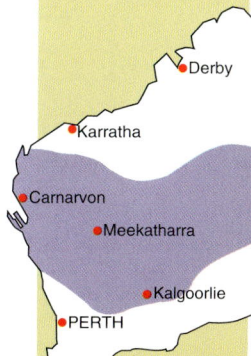

Almost mature winged seed, about actual size. Note the broad, flattened leaves

BLUEBUSH GROUP

Shy bluebush

Maireana platycarpa

Family: Chenopodiaceae

GENERAL DESCRIPTION: Shy bluebush is a brittle, low, many-branched, perennial shrub, which grows to 60 cm in height and up to 1 m in diameter. The woody basal branches lie horizontally and the younger, slender branches grow vertically from the old wood. The leaves are 7–12 mm long and are oval to flattish in cross-section. They are succulent and covered in a dense layer of white to grey hairs. Shy bluebush flowers in winter and the fruits mature in September or October. The fruit consists of a woody seed case about 5 mm in diameter from which protrudes a single vertical process up to 6 mm long. The case itself is surrounded by a pink to brown wing, up to 2 cm in diameter. The wing has a sparse covering of short, white hairs.

Shy bluebush, about 80 cm across

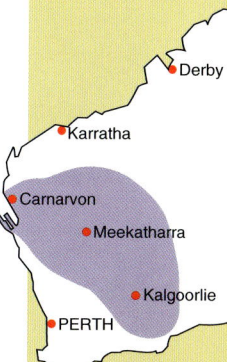

HABITAT: Shy bluebush grows in association with other saltbush pasture plants in alkaline soils on floodplains, valley floors and salt lake margins. It is particularly common in the coastal Gascoyne where it occurs with Gascoyne bluebush. It also occurs in other habitats with bladder saltbush and river saltbush.

FORAGE VALUE: It is very palatable to stock and the foliage persists during droughts. The crude protein content varies from 6 to 16 per cent while digestibility ranges from 52 to 66 per cent.

INDICATOR VALUE: Shy bluebush is an indicator of good range condition where it is common. It is removed by overuse or injudicious stocking and its absence from those pastures where it should occur indicates poor range condition. Because it is infrequent in the inland, its indicator value there is limited.

Fruit, about half actual size

BLUEBUSH GROUP

Gascoyne bluebush

Maireana polypterygia

Family: Chenopodiaceae

GENERAL DESCRIPTION: Gascoyne bluebush is a compact shrub growing to a height of 1 m and to 1.5 m across. Many individuals growing together in a clump can give the appearance of a much larger plant. The leaves are fleshy and a purplish green, particularly when growing vigorously. They are sub-round, some being flattened on the upper surface and rounded below. They have a covering of short hairs. They tend to be a dull green when conditions are dry. The flowers are insignificant and are followed by seed cases which are about 1 cm in diameter. The fruits have two horizontal wavy wings, one at the top and the other at the bottom of the seed cases. Imperfect wings, running vertically, connect the upper and lower wings.

HABITAT: Gascoyne bluebush grows on alluvial plains and on sluggish, saline drainage systems in stony uplands. The soils are usually alkaline loams and clays and are frequently saline. It grows with shy bluebush, wait-a-while and snakewood.

Gascoyne bluebush, about 1.3 m across

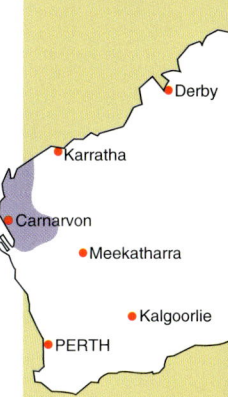

FORAGE VALUE: Gascoyne bluebush is high in salt. It is readily eaten when fresh water is available, but is ignored if the water is brackish. It is resistant to grazing and is moderately tolerant of drought conditions. The crude protein varies from 15 to 19 per cent and the digestibility from 60 to 74 per cent.

INDICATOR VALUE: It is a hardy species, tolerant of heavy grazing. It is not a good indicator of range condition. Shy bluebush should, in most cases, grow with it. If areas with Gascoyne bluebush lack shy bluebush, they are in fair to poor range condition. If Gascoyne bluebush is the only perennial plant present and it occurs only in scattered clumps or as isolated plants, then the pasture is in poor range condition.

Succulent leaves and black seed cases, about one and a half times actual size

BLUEBUSH GROUP

Sago bush   Black bluebush

Maireana pyramidata

Family: Chenopodiaceae

GENERAL DESCRIPTION: Sago bush is a greyish green shrub which grows to 2 m, but is commonly less. It has stiff, woody, angular branches and forms a compact bush. The leaves are fleshy, less than 5 mm long and round in cross-section. They are a distinctive bluish green colour and are sparsely covered in very fine, short hairs. The flowers are very small and inconspicuous. The seed is enclosed in a large, conical, but not woody seed case, surrounded by a membranous wavy wing, above which is a small cone or pyramid. When mature, the fruits turn brown but they are often tinged with pink when young. While most bluebush species have bisexual flowers, sago bush differs in that it bears separate male and female flowers. In some instances there are separate male and female plants.

HABITAT: Sago bush grows in a wide range of habitats including river floodplains, the margins of salt lakes, plains below breakaways, basaltic areas,

Sago bush, about 70 cm across

terraces of gilgais on saline upland drainages and on limestones. The soils are alkaline sands over clays, or clays with crabholes. It grows with river saltbush, golden bluebush and snakewood.

FORAGE VALUE: It is not a highly preferred feed and other species are eaten before it. It is drought resistant, withstands heavy grazing well and is a valuable component of the pasture. The leaves contain up to 22 per cent crude protein and have a digestibility of about 60 per cent.

INDICATOR VALUE: Sago bush is not a reliable indicator of range condition, since it is resistant to both drought and grazing. When the vigour of sago bush declines, it indicates severe overuse. An increase in the number of plants can indicate a conservative stocking history.

REHABILITATION: Sago bush is an excellent addition to perennial shrub seeding mixes. The seed may be collected locally and it appears to germinate easily. It tolerates alkaline and saline conditions and is a vigorous coloniser. The seeding rate should be about 1 kg/ha in a much larger mix.

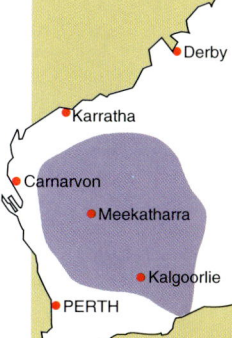

Winged fruit with central 'pyramid' and succulent leaves, about one and a half times actual size

BLUEBUSH GROUP

Pearl bluebush

Maireana sedifolia

Family: Chenopodiaceae

GENERAL DESCRIPTION: Pearl bluebush is a compact, bluish grey shrub with tough, upright, woody stems which grows to 1.5 m. It often forms dense clumps, which consist of many individuals. The leaves are 4–8 mm long and succulent. They are round in cross-section, have rounded tips and are covered in very short, white hairs. There are separate male and female plants, and the flowers are borne in late spring in the axil between leaf and stem. The flowers commonly follow late winter rain and mature in December. Pearl bluebush, however, is irregular in flowering habit and does not produce seed after each late winter rainfall event. The single seed is enclosed in an oval seed case, 3 mm in diameter. It is surrounded by a fine, straw-coloured wing, 10 mm in diameter.

Pearl bluebush, about 1.7 m across

HABITAT: Pearl bluebush grows on undulating plains with relatively deep, alkaline loams or clays with or without an upper storey. It grows with bladder saltbush, pink-seeded bluebush, sheoaks, salmon gums and mulga.

FORAGE VALUE: Pearl bluebush contains up to 20 per cent crude protein and 10 per cent salt and is eaten by stock if fresh water is available.

INDICATOR VALUE: It is an extremely hardy plant and is reputed to live up to 300 years. It is not a sensitive indicator of change in range condition. New plants are rare and should be fostered by management. It can recover from burning, but large areas of the Nullarbor Plain have been stripped of this plant by the combined effects of repeated bush fires and grazing by rabbits. Populations decline under heavy grazing and its absence from an area which should support it indicates poor range condition.

Stems and succulent leaves covered with grey, woolly hairs, about one and a half times actual size

BLUEBUSH GROUP

Lax bluebush

Maireana thesioides

Family: Chenopodiaceae

GENERAL DESCRIPTION: Lax bluebush is a sparse, scrambling shrub, growing to 1.5 m or more. It has long, drooping branches and can be up to 2 m across. The leaves are dark green with a reddish tinge and are up to 1.5 cm long. They have a pointed tip, are round in cross-section, succulent and hairless. The flowers are very small and are succeeded by a seed case up to 6 mm in diameter. It is surrounded by a flat, hairless, membranous wing up to 1.5 cm wide. The seed cases become black on maturity.

HABITAT: Lax bluebush grows on a wide variety of habitats including floodplains, sand banks, the margins of saline areas and stony hills. It is found in association with larger shrubs and trees such as mulga, curara, sago bush and Wilcox bush.

Lax bluebush, about 90 cm across

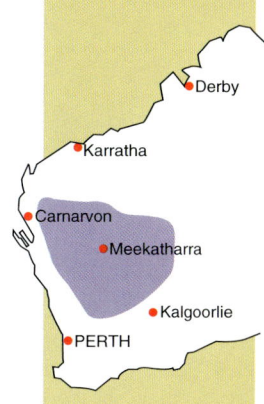

FORAGE VALUE: Lax bluebush is extremely attractive to stock. It is never profuse and, therefore, produces insignificant quantities of forage. It is, however, preferentially grazed when the more palatable of the annual species are consumed. It drops its leaves during severe drought and cannot be regarded as a good stand-by feed.

INDICATOR VALUE: It is a decreaser species and its presence indicates good range condition. It is always absent in heavily grazed areas.

Hairless fruit and leaves, about one and a half times actual size

BLUEBUSH GROUP

Felty bluebush

Maireana tomentosa

Family: Chenopodiaceae

GENERAL DESCRIPTION: Felty bluebush is an erect and compact low shrub, usually less than 50 cm high. The leaves are alternately arranged along the stem and are up to 12 mm long and 2 mm in diameter, sub-circular in cross-section, succulent and light green or bluish green. Both leaves and stems have a dense covering of woolly, golden hairs and the leaves are frequently broader towards the tip. The flowers are inconspicuous. The seed cases are up to 2 mm in diameter and are surrounded by a flat, translucent, gold-coloured wing up to 1 cm in diameter. The plant is variable in its appearance, both the vegetative growth and the fruits differing from site to site. It is not known whether this is due to growth conditions or to genetic differences.

Felty bluebush, about 20 cm high

HABITAT: Felty bluebush occurs as an under-storey plant in a wide range of habitats ranging from floodplains to mulga woodlands. It grows with Gascoyne bluebush, shy bluebush, bladder saltbush, tall saltbush and mulga.

FORAGE VALUE: Felty bluebush is a valuable forage containing up to 24 per cent crude protein. It has a digestibility of up to 64 per cent. It quickly falls away in droughts, but is quick to respond to rain.

INDICATOR VALUE: It is a relatively short-lived species, common in good seasons. It is a useful guide to the level of use of the current season's production, but has no value in indicating changes in range condition.

Fruit with a slit in the wing, about twice actual size. Note the felty hairs on the leaves

BLUEBUSH GROUP

Pink-seeded bluebush   Downy bluebush

Maireana trichoptera

Family: Chenopodiaceae

GENERAL DESCRIPTION: Pink-seeded bluebush is a weak perennial, which grows to 30 cm. The branches are woolly and very slender. The leaves are 5–10 mm long, sub-round in cross-section, about 1.5–2 mm across, fleshy and covered with fine, white hairs. The inconspicuous small flowers occur between the leaf and stem in leafy spikes and produce a downy fruiting body in which the membranous wing is edged with pink. The seed is held in a cylindrical seed case in the centre of the wing. It is a common and easily recognised bluebush with an erect habit, woolly covering and pink-edged seed cases crowded along the ends of the branches.

HABITAT: Pink-seeded bluebush is found on calcareous platforms or calcareous plains with sandy or loamy soils with species common in bluebush and saltbush communities.

Pink-seeded bluebush, about 30 cm across

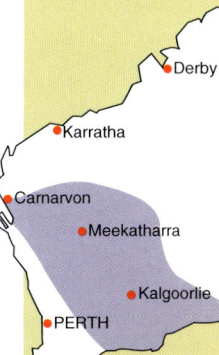

FORAGE VALUE: It is readily eaten by stock, particularly in the absence of other more acceptable plants. It contains up to 16 per cent crude protein and 10 per cent salt.

INDICATOR VALUE: It is a short-lived perennial and an important component of a pasture. Pastures should be managed to maintain it and other edible low bluebushes and bindiis. It is an indicator of good range condition if it is found in abundance. It is also a useful guide to the level of use of the current season's production.

REHABILITATION: Pink-seeded bluebush is a useful colonising plant since it seeds prolifically. Seed should be collected locally, dried thoroughly and stored in a dry atmosphere.

Woolly hairs covering fruit, stem and leaves, about actual size

BLUEBUSH GROUP

Three-winged bluebush

Maireana triptera

Family: Chenopodiaceae

GENERAL DESCRIPTION: Three-winged bluebush is a dark green, upright, dense shrub growing to 60 cm. The leaves are hairless, bright green, succulent, about 1 cm long and round in cross-section. They are arranged densely along the branches. The flowers are small and insignificant and are followed by abundant and prominent fruits which are green or red initially, turning black at maturity. The fruits have three membranous, wing-like appendages attached to the seed case. The plant appears black when it is covered with mature fruits. Leaves under moisture stress also turn black.

HABITAT: Three-winged bluebush grows in saline pastures, in crabholes on stony slopes and on other alkaline and sub-saline areas receiving run-on water. It is found with golden bluebush, sago bush, saltbush and snakewood.

FORAGE VALUE: It is not very drought tolerant, and is less palatable than most bluebushes, but is eaten in some situations. It contains up to 18 per cent crude protein when actively growing.

Three-winged bluebush, about 60 cm across

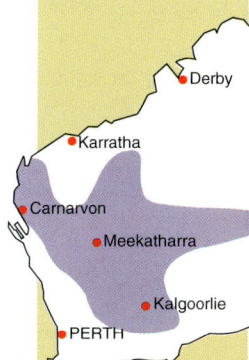

INDICATOR VALUE: It is not a reliable indicator of range condition since it occupies a different position from one environment to another. In saltbush pastures and in crabhole communities an abundance of three-winged bluebush indicates poor range condition, while it is a normal component of pastures below breakaways and in some sluggish drainages. In these situations it may be a useful guide to changes in range condition, though judgements must be tempered by the knowledge that it is not drought tolerant. Changes in the number of three-winged bluebush may be only a reflection of the current seasonal conditions.

REHABILITATION: Three-winged bluebush is common around many mining sites and can be used in a seeding mix of perennials. It will colonise difficult alkaline and saline sites and does not compete with more persistent and durable perennials.

The black seeds are mature while the yellow seeds are immature, about three-quarters actual size. Note the wing is divided into three parts

Lignum

Muehlenbeckia cunninghamii

Family: Polygonaceae

GENERAL DESCRIPTION: Lignum is an upright shrub, growing to 2 m. It appears to be leafless. It has an angular, very tangled and intricate branching habit. The short side branchlets usually end in a spine. The leaves are usually absent on older branches and are found scattered on young stems only. The very small, pale yellow flowers are borne in clusters along the branches and at their ends. It flowers for most of the year. The seed is carried in a cone-shaped structure up to 5 mm long.

Lignum, about 1 m high

HABITAT: Lignum is entirely restricted to non-saline claypans, swamps and crabholes subject to long periods of flooding. It grows in closed communities with few other plants.

FORAGE VALUE: It is rarely eaten and the forage value is correspondingly very low.

INDICATOR VALUE: It has no indicator value.

Flowers, about actual size

Nitre bush   Dillon bush

Nitraria billardierei

Family: Zygophyllaceae

GENERAL DESCRIPTION: Nitre bush is a rigid and spreading shrub, which can reach 2 m but is commonly about 1 m high. The smaller branches end in a spike. The leaves are thick and fleshy, up to 3 cm long and 7 mm wide, and are covered with fine hairs. They are often a pale yellow, are distributed irregularly along the stems or may be clustered in dense groups. The flowers are small, white, with five petals and are borne in small clusters along the branches. The fruit is purple and fleshy. It is up to 2 cm long and contains a hardened seed case.

HABITAT: Nitre bush has two distinct habitats. It grows prolifically along sand dunes in the Gascoyne River region and along the Eucla coast. It is sporadically recorded on saline soils in the interior where it grows in moisture-favoured locations with saltbush and bluebush.

Nitre bush, about 1 m high and 4 m across

FORAGE VALUE: It is not highly palatable and is used only sparingly by sheep.

INDICATOR VALUE: It has no indicator value.

REHABILITATION: Nitre bush can be used as a coloniser of waste dumps where surface conditions permit the accumulation of water. Seed supplies should be readily available, but the fleshy part of the fruit must be removed or dried completely before storage.

Nitre bush fruits, about actual size

Goldfields daisy

Olearia muelleri

Family: Asteraceae

GENERAL DESCRIPTION: Goldfields daisy is a small, many-branched shrub which grows to 60 cm. The dark green leaves are about 1 cm long and 4 mm wide, tending towards broadness at the tips and arranged alternately along the stem. The showy flowers are white, with 10–12 petals, each being about 1.5 cm long. The centre of the flower containing the seeds is yellow. The seeds resemble those of a thistle.

Goldfields daisy, about 50 cm high

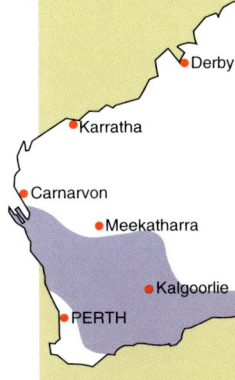

HABITAT: Goldfields daisy is found in alkaline and sandy loams with gums and other shrubs common in the Goldfields gum belt.

FORAGE VALUE: Goldfields daisy has little forage value.

INDICATOR VALUE: It has no indicator value.

A flower, about actual size

Seed being released, about actual size

Shrubby rice flower   Banjine

Pimelea microcephala

Family: Thymelaeaceae

GENERAL DESCRIPTION: Shrubby rice flower is a tall, very open, often straggly shrub which grows up to 2 m. The leaves are soft green, narrow and oblong in shape, and up to 3 cm long and 4 mm wide. They are arranged opposite each other on quite short branchlets off the larger stems. There are separate male and female shrubby rice flower plants. The flowers in both sexes are borne in clusters and are usually pale cream. The female flower is a very narrow tube about 5 mm long, which has four conspicuous 'petals' at the end. They are gathered into groups of up to 12 at the ends of branches. The male flowers are shorter and in groups of up to 100. The reddish or yellow fruit is about 4 mm long and contains one seed. It is succulent initially, but dries at maturity.

Shrubby rice flower, about 1 m high

HABITAT: Shrubby rice flower is found on sandy and sandy loam soils on sand dunes and low sandy banks where it occurs with various *Acacia* species such as wanyu and mulga. It is also found more rarely on alluvial plains and limestone banks where it occurs in slightly saline areas with low mulga, prickly acacia and saltbush.

FORAGE VALUE: It has low stock palatability. It is also extremely drought resistant. Other *Pimelea* species have been incriminated as the cause of stock mortality in the eastern states of Australia and shrubby rice flower has also been suspected of poisoning stock.

INDICATOR VALUE: It has no indicator value.

Flowers, about actual size

A berry, about actual size

Low mulla mulla

Ptilotus beardii

Family: Amaranthaceae

GENERAL DESCRIPTION: Low mulla mulla is a prostrate, spreading shrub, somewhat like a frankenia in growth habit, but lacking the salt crystals on the leaves. It rarely exceeds 40 cm in height and is usually 50–80 cm across. The lower branches are very gnarled while the new growth is spindly and held erect on the new wood. The leaves are greyish green, 2–3 mm long, hairy and pointed. The flowers occur at the ends of the new growth in open, globular spikes each about 1 cm across. Each flower is about 2 cm long and is pinkish or purple at the tips, tending to white at the base.

HABITAT: It occurs on stony slopes beneath breakaways, on alkaline stony soils with sago bush and on saline lake fringes.

Low mulla mulla, about 30 cm high

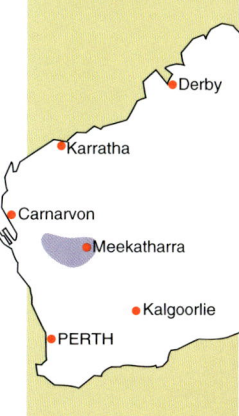

FORAGE VALUE: Low mulla mulla is readily eaten and has fair to good drought tolerance. It makes rapid growth after rain and is very resistant to grazing.

INDICATOR VALUE: Low mulla mulla is an important shrub which gives stability to land which is preferentially grazed. It is an indicator of fair to good range condition.

Low mulla mulla flowers and leaves, about one third actual size

Narrow-leaved mulla mulla   Climbing mulla mulla

Ptilotus divaricatus

Family: Amaranthaceae

GENERAL DESCRIPTION: Narrow-leaved mulla mulla is a dark green, dense and practically hairless shrub growing to 1.5 m. It is often found intertwined with other shrubs and trees. The leaves are narrow, dark green, shiny and up to 3 cm long and 5 mm wide. The flower heads are white and globular and have a shape similar to cotton bush, but they are less dense. They are borne singly at the end of each branch.

HABITAT: Narrow-leaved mulla mulla is common on saline loam and clay soils which are subject to flooding. It grows with mulga, sago bush, saltbush and sage.

Narrow-leaved mulla mulla, about 80 cm high

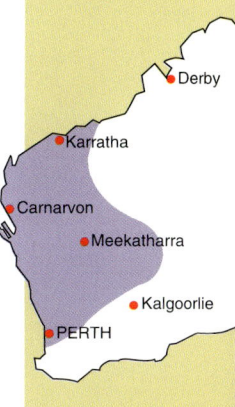

FORAGE VALUE: It is highly palatable and is removed preferentially by stock. Continued overuse will lead to its eradication from the pasture. It is drought tolerant and responds quickly to rainfall. As most edible material is within reach of stock, it is a valuable species and should be encouraged.

INDICATOR VALUE: It is a decreaser species and is an indicator of good range condition. It is common in saltbush pastures in good condition where it will be found growing apart from other shrubs and not beneath their protective canopy as it does when range condition is only fair.

A flower head, about actual size

Saline mulla mulla

Ptilotus lazaridis

Family: Amaranthaceae

GENERAL DESCRIPTION: Saline mulla mulla is a loose, scrambling shrub, which grows to 1 m in height and diameter. It has brittle stems bearing narrow, bright green to yellow leaves and pink flower heads. The leaves are up to 2 cm long and 3 mm wide. The pink flowers occur in small, open balls up to 1 cm in diameter and are borne near the ground at the end of long slender stems. It can develop a massive rootstock under heavy grazing and this shoots vigorously after rain.

HABITAT: Saline mulla mulla occurs on saline plains which are subject to flooding. It is often intertwined with more robust shrubs such as silver

Saline mulla mulla, about 30 cm high

saltbush, swamp saltbush or fuchsia bush, but grows independently of other shrubs in moderately grazed situations.

FORAGE VALUE: It is very palatable to sheep and cattle. It is drought tolerant and resistant to grazing.

INDICATOR VALUE: It is an indicator of good range condition. Populations decrease with heavy grazing.

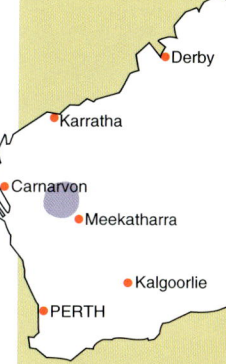

Flower heads, about actual size

Cotton bush

Ptilotus obovatus

Family: Amaranthaceae

GENERAL DESCRIPTION: Cotton bush is a dense, greyish shrub growing to 60 cm. It has multiple stems which originate at ground level, giving it a generally compact appearance. It flowers in September and October and is then covered with a mass of white to pink flower heads. The leaves are greyish, 2 cm long and 1 cm wide, with a dense, felty covering of fine hairs. Each branch produces a number of globular flower heads, each up to 2 cm in diameter.

HABITAT: Cotton bush is found in all habitats except those which are saline, or subject to frequent inundation. It occurs on shallow loams underlain by hardpan, on sands or on skeletal soils on rock. With mulga, it is one of the commonest plants in the shrublands.

FORAGE VALUE: The crude protein content of cotton bush ranges from 8 to 14 per cent and the digestibility from 41 to 58 per cent. It is palatable to stock,

Cotton bush, about 40 cm high

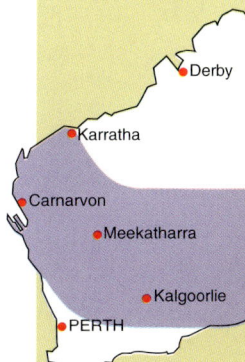

but drops its leaves when under water stress. However, it has value during droughts through its ability to respond quickly to minor rainfall events, which are otherwise inadequate to induce annuals to grow. Plants in full flower can be deceptive, as a close examination will show that most of the leaves have been shed at this stage. In some instances it appears to be unpalatable, but this could be a varietal difference.

INDICATOR VALUE: Populations of cotton bush fluctuate a great deal in response to rainfall and some care has to be taken when using it as an indicator of range condition. It is a useful plant in the inland mulga shrublands, but of doubtful value in the coastal Gascoyne where it replaces the more valuable bluebushes and saltbushes.

REHABILITATION: Cotton bush germinates very poorly. Although seed can be gathered easily, it should never be a major component in a seed mix at the expense of other species, because of its low germination rate. Cotton bush seed should be applied in addition to other seeds.

Flower heads, about actual size

Gascoyne mulla mulla

Ptilotus polakii

Family: Amaranthaceae

GENERAL DESCRIPTION: Gascoyne mulla mulla is a shrub which can be 40 cm high and about 75 cm across. The principal branches are twisted, stout and woody. The new growth arising from them is fragile and erect. The leaves are a bright green, about 1 cm long and 3 mm wide and broadly elliptical in shape. The pink flowers are borne at the ends of the stems. They are in very open, sub-globular spikes, each flower in the spike being about 2 cm long and clothed in short, silky, pinkish hairs.

HABITAT: Gascoyne mulla mulla grows on the inter-banks of the low sand bank country in the west Gascoyne. It is thought to have replaced Gascoyne bluebush in this habitat.

Gascoyne mulla mulla, about 30 cm high

FORAGE VALUE: Gascoyne mulla mulla is not relished by stock but it is an important shrub in the sand bank/inter-bank communities of the Gascoyne River and delta and the Carnarvon coastal plain. It tends to drop its leaves early in dry periods, but responds quickly to small amounts of rainfall. It appears to be very long-lived and resistant to grazing pressure.

INDICATOR VALUE: It has little indicator value since it is very resistant to both drought and overuse. Inter-bank vegetation in the Gascoyne lacking Gascoyne mulla mulla and dominated instead by plants such as needle bush is in very poor range condition. Equally, pastures in which Gascoyne mulla mulla is the only species present could be said to be overused. Such areas usually have windblown soil piled around the bases of the shrubs, which is a characteristic of overuse.

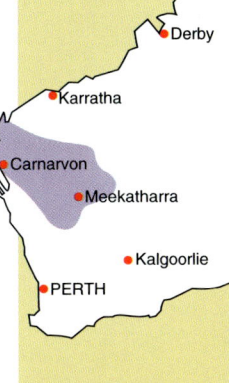

Flower heads, about three-quarters actual size

Royal mulla mulla

Ptilotus rotundifolius

Family: Amaranthaceae

GENERAL DESCRIPTION: Royal mulla mulla is a robust, perennial shrub which grows to 2 m. It has upright stems and roundish, grey leaves. The flower heads are long, pink spikes, held in a candelabra form. They are very prominent and make the plant conspicuous on the stony plains and rocky hillslopes it favours. The leaves have wavy edges and are up to 6 cm in diameter. They are densely covered with white hairs. The flower heads are 10 cm long and 4 cm wide and have a sweet, cloying odour. The flowers persist for many months following the flowering season and gradually fade from pink to white, the upper flowers in the head usually shedding first to expose the white flower stem.

Royal mulla mulla, about 1 m high

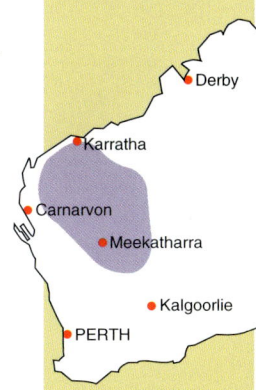

HABITAT: Royal mulla mulla is found mainly on shallow, stony slopes underlain by basalts, granites, metamorphic rocks or hardpan. It grows with mulga, *Calytrix* and *Eremophila* species.

FORAGE VALUE: Its forage value is unknown. It is occasionally eaten by stock.

INDICATOR VALUE: Royal mulla mulla has no known indicator value.

A flower head, about three-quarters actual size

Horse mulla mulla

Ptilotus schwartzii

Family: Amaranthaceae

GENERAL DESCRIPTION: Horse mulla mulla is an herbaceous perennial with a large, woody, above-ground butt. It has many upright, almost leafless stems up to 1 m high, bearing pink flowers at their tips. The few narrow leaves on this plant are up to 4 cm long and grow from the base. The pink flowers are held in tight globular heads. On maturity they fade to white and drop.

Horse mulla mulla, about 70 cm high

HABITAT: Horse mulla mulla is found in sparse stands growing on shallow soils above hardpan. It is also common on pebble-strewn, rocky slopes throughout the shrublands.

FORAGE VALUE: It is eaten by sheep, cattle and especially by horses, but is relatively unimportant as it produces only a small amount of edible material.

INDICATOR VALUE: It is a decreaser species and its presence indicates good or fair condition.

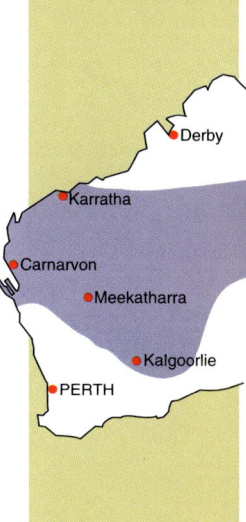

Leafless stem and flower head, about twice actual size

Lake-fringe rhagodia Low rhagodia

Rhagodia drummondii

Family: Chenopodiaceae

GENERAL DESCRIPTION: Lake-fringe rhagodia is a globular, green-grey shrub that grows up to 1.5 m high and 3 m wide. The branches root where they come in contact with the soil. The leaves are fleshy, spear-shaped and up to 10 mm long, 3 mm wide and have a covering of mealy hairs. There are separate male and female plants. The flowers on each are very small and held in terminal spikes, which also have a dense covering of mealy hairs. The female flowers produce 2 mm diameter seeds with a fleshy, red outer layer.

Lake-fringe rhagodia, about 1.5 m across

HABITAT: Lake-fringe rhagodia grows on sandy rises adjacent to salt lake margins beneath a canopy of mulga or eucalypts.

FORAGE VALUE: It is readily eaten by stock. It holds its leaves well during dry periods and is a good drought reserve. It is resistant to grazing.

INDICATOR VALUE: It is a decreaser species. Its presence indicates good range condition.

Leaves and stems, about half actual size

Tall saltbush Thorny saltbush

Rhagodia eremaea

Family: Chenopodiaceae

GENERAL DESCRIPTION: Tall saltbush is a grey, straggly, and many-branched shrub growing to 2 m. The leaves have a musty smell when crushed, are spear-shaped and often lobed at the base. They can be up to 2 cm long and 1 cm wide. They have a covering of grey, waxy hairs. Male and female flowers are on separate shrubs and form upright clusters. The female flowers produce bunches of red, succulent berries up to 4 mm in diameter.

HABITAT: Tall saltbush occurs on a great variety of soil types and grows with mulga, curara, prickly acacia, cotton bush and Wilcox bush. On the Ashburton floodplain, where it appears to be an increaser species, it can form dense, almost impenetrable stands.

Tall saltbush, about 1.3 m high

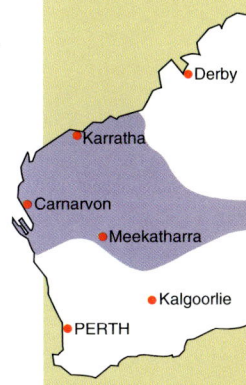

FORAGE VALUE: It is moderately palatable to stock but loses its leaves during dry periods so that only dry, spiny stem material is available. The crude protein content of the leaf varies from 6 to 12 per cent and digestibility can be as high as 67 per cent. It contains little salt.

INDICATOR VALUE: It is extremely resistant to grazing and for this reason it is not a reliable indicator of trend in range condition. More sensitive species in the community being examined should be used for the assessment of change.

REHABILITATION: Tall saltbush could be included in a seed mix of perennial species if seed can be gathered locally. It is unlikely to be successful on alkaline or saline sites as it is found on acid soils under natural conditions.

Male flowers, about actual size

Berries, about three-quarters actual size

Caustic bush

Sarcostemma viminale subsp. *australe*

Family: Asclepiadaceae

GENERAL DESCRIPTION: Caustic bush is a leafless, perennial shrub growing to 2 m and to 2.5 m wide. The stems are stout, 5–10 mm in diameter and may be up to 1 m long. When damaged they exude a caustic, milky sap. The flowers are waxy, creamy white and star-shaped. They are held in stalkless clusters of two to eight flowers. The fruit consists of narrow cylindrical pods 5–14 cm long, which are pointed at the tip. They split lengthwise when ripe, to release numerous seeds, each with a tuft of silky hairs at one end. The seed is wind-distributed.

Caustic bush, about 1.5 m across

HABITAT: It is never common in any situation and appears to be distributed throughout the shrublands. It is not found on sandplains.

FORAGE VALUE: Caustic bush is a confirmed toxic plant in eastern Australia, causing death in sheep, cattle and horses. It has not been incriminated in Western Australia, though it is consumed readily enough and has largely disappeared in most communities.

INDICATOR VALUE: Caustic bush has no indicator value.

Currant bush   Maroon bush

Scaevola spinescens

Family: Goodeniaceae

GENERAL DESCRIPTION: Currant bush is a variable shrub which grows to 3 m under favourable conditions. Soil type and salinity appear to influence its growth habit. It is many-branched and intricate in appearance. The leaves are a light green, more or less oblong, 1–3 cm long and clustered together in bundles along the stems. Plants from non-saline habitats tend to have brighter and larger leaves, while the leaves of those from saline environments are small and dull in appearance. Each bundle of leaves has an associated small spine about 1 cm long. The flowers are fan-shaped, usually white or pale cream and they may be streaked with purple or bronze markings. The fruits are black berries about 5 mm in diameter.

Currant bush, about 80 cm high

HABITAT: Currant bush is found in a wide range of habitats. It is common in saltbush pastures and also occurs on rocky slopes, mulga plains and on low sand banks surrounding bluebush and saltbush pastures. It is also found in sandplain country beneath wanyu and bowgada.

FORAGE VALUE: Currant bush is an important forage plant. The leaves contain up to 13 per cent crude protein and are relished by stock. It is drought resistant and retains its leaves well into drought periods. It is a protected plant in the Goldfields where, under the name maroon bush, it is collected to make infusions to ease pain in cases of abdominal disease.

INDICATOR VALUE: The presence of vigorous currant bush indicates good range condition. It rarely dominates an area and is slow to re-establish, so these areas should be given prolonged freedom from use. It also appears to be a slow grower from the seedling stage.

Mature and immature fruit, about three-quarters actual size

Flowers, about three-quarters actual size

341

Ragged-leaf fan flower   Felted scaevola, Grapevine bush

Scaevola tomentosa

Family: Goodeniaceae

GENERAL DESCRIPTION: Ragged-leaf fan flower is a scrambling perennial growing to 1.5 m. The leaves are yellowish green and spear-shaped, with about four pointed lobes on each side. The leaves, which can be up to 3 cm long and 1.5 cm wide, have a covering of fine, white hairs. The orange, fan-shaped flowers are about 5 mm wide. The petals are black on the underside and narrow towards the base, to form a tube about 1.5 cm long. The fruit is a black berry about 7 mm in diameter.

HABITAT: Ragged-leaf fan flower grows on sandy banks and on calcrete platforms which have sandy or loamy soils. It is found with Wilcox bush, tall poverty bush, soft spinifex and limestone wattle.

Ragged-leaf fan flower, about 30 cm high

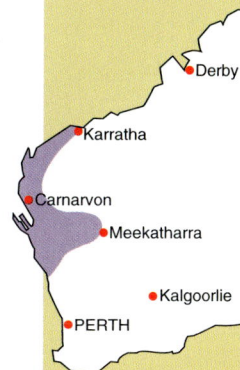

FORAGE VALUE: It is a palatable species that is eaten when annual feed is scarce. It is not resistant to drought and drops its leaves at the onset of severe moisture stress. The crude protein content varies from 7 to 14 per cent while the digestibility ranges from 52 to 67 per cent.

INDICATOR VALUE: Its presence indicates good or fair range condition.

Flowers and immature fruit, about actual size

Giant bindii

Sclerolaena hostilis

Family: Chenopodiaceae

GENERAL DESCRIPTION: Giant bindii is a dense, globe-shaped, light green perennial shrub growing to 1.3 m. The leaves are a light, shiny green, 1 cm long, succulent and hairless. The woody seed cases are attached firmly to the branchlets and are about 1 cm in diameter. Four large, stout, woody spines, 1 cm long, protrude from the top of each seed case.

HABITAT: Giant bindii grows on semi-saline clays with Roebourne Plains grass on the coastal margins of the Roebourne Plain. It also grows in clay valleys with WA buffel grass in the granite massif of the central Pilbara.

Giant bindii, about 1.5 m across and 1 m high

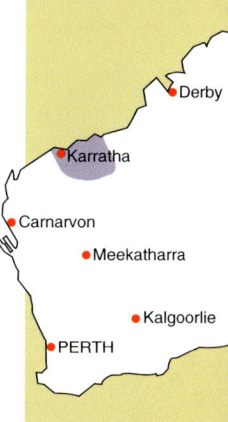

FORAGE VALUE: It is readily eaten by stock.

INDICATOR VALUE: This species indicates good range condition.

NOTE: It appears that the seeds of this species are only released when the shrub dies. A dramatic example of its dispersion occurred during cyclone Ilona in 1989, when large giant bindiis were broken up and small branches scattered over a wide area.

Seed capsules and leaves, about half actual size

Showy groundsel

Senecio magnificus

Family: Asteraceae

GENERAL DESCRIPTION: Showy groundsel is an erect, hairless, woody, perennial shrub which grows to about 1 m. The leaves are green and fleshy, often grasping the stem, and the ends of the lower ones are notched. The flower heads are grouped at the ends of the branches. Showy groundsel belongs to the group of plants in this family which has showy petals. The petals are bright yellow but sparse, rarely exceeding eight or ten. They are about 15 mm long and surround a disc of true flowers which is about 1 cm across.

HABITAT: Showy groundsel is distributed sparsely in the arid shrubland zone, where it is a weed colonising severely degraded areas. It is well established on

Showy groundsel, about 1.2 m high

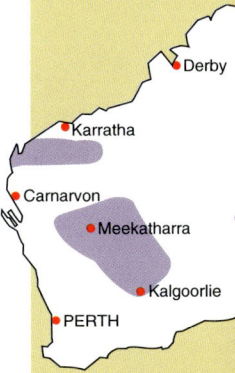

creeklines north of Leonora and is infrequent elsewhere on roadsides and in watercourses. It avoids saline sites and is associated with mulga communities.

FORAGE VALUE: It has no forage value. Some cases of stock poisoning are associated with this plant in the Northern Territory.

INDICATOR VALUE: It denotes poor range condition.

Flowers, about three-quarters actual size

Tall sida

Sida calyxhymenia

Family: Malvaceae

GENERAL DESCRIPTION: Tall sida is an erect shrub with fine, branched hairs on the leaves and stems, which give it a somewhat coarse appearance. It is a many-branched shrub with numerous short stems and reaches 2 m in height. The leaves are mid-green on the upper surface and pale green below. They are broadly toothed on the margins, up to 2 cm long and almost 1 cm across. The flowers are borne singly with five distinct, yellow petals. They are up to 1 cm across and are supported by broad, membranous sepals which expand after the petals have dropped and persist with the fruit, which is sub-globular to sub-conical in shape and about 6 mm in diameter.

Tall sida, about 40 cm high

HABITAT: Tall sida is found in communities dominated by mulga. It is found on hillsides, in creeklines, in watercourses on alluvial plains and in mulga thickets. It avoids soils which are saline. It is a common member of most mulga communities.

FORAGE VALUE: It is palatable to stock and moderately drought resistant. It is, therefore, an important element of shrub pastures beneath the mulga cover and should be encouraged.

INDICATOR VALUE: Tall sida populations decline in heavily grazed situations. When it is present as vigorous, well-grown bushes in a community, it indicates good range condition. When it is present only as stunted and obviously heavily grazed plants, it indicates overgrazing.

Flowers, about one and a half times actual size

Prostrate sida   Dwarf sida

Sida corrugata

Family: Malvaceae

GENERAL DESCRIPTION: Prostrate sida is a prostrate, short-lived, herbaceous shrub which grows to 15 cm high and 50 cm wide. The leaves are up to 1.5 cm long and 1 cm wide and have shallow, notched edges. The undersides of the leaves and the stems are grey, due to a hairy covering, while the upper leaf surfaces lack the hairs and are dark green. The flowers are up to 8 mm wide with five separate yellow petals. The flowers mature to form small, round capsules about 5 mm in diameter which split open to release the seeds.

HABITAT: Prostrate sida occurs on valley floors with loamy or cracking clay soils and grows with bladder saltbush, bluebush and sago bush.

Prostrate sida, about 15 cm high

FORAGE VALUE: It is readily eaten by stock, but as leaves fall off with moisture stress, it does not provide feed in droughts. However, it quickly produces new growth with falls of rain which are insufficient to cause the germination and establishment of annuals.

INDICATOR VALUE: It tends to increase when the vigour of shrub competition is reduced by heavy grazing or fire. Dense stands indicate that the pasture composition may have been altered. The change may not necessarily be detrimental, since dense prostrate sida stands are productive. Land covered by prostrate sida may lack stability and this could be a cause for concern, since during droughts and at the break of droughts, the surface soil would have little protection and could be damaged. It is, however, a useful guide to the current level of use.

Flower and immature fruit, about actual size

Flannel bush

Solanum lasiophyllum

Family: Solanaceae

GENERAL DESCRIPTION: Flannel bush is an herbaceous perennial with a woody base. It can grow to 1 m but is commonly about 60 cm high. It has strong, densely hairy, but not woody branches armed with stiff, short spines up to 1 cm long. The leaves are grey, up to 10 cm long and 6 cm wide, with a wavy margin. They are thick and densely covered with fine hairs. The prominent veins beneath the leaves have short spines. The flowers are bright purple, 2 cm across and consist of five united petals above which rise prominent, yellow, united stamens. The fruit is a woolly berry about 1 cm in diameter.

HABITAT: Flannel bush is found in most rangeland types in the shrublands, avoiding only saline saltbush communities.

FORAGE VALUE: Flannel bush is palatable and is heavily grazed, particularly when young, before the plant becomes rigid. In periods of feed shortage even

Flannel bush, about 50 cm high

mature leaves and slender stems may be eaten. The leaves contain up to 15 per cent crude protein. It avoids severe moisture stress by dropping its leaves. It responds quickly to small amounts of rain, and will provide grazing after rains which are insufficient for the germination and establishment of annuals.

INDICATOR VALUE: Flannel bush indicates poor range condition when found dominating a pasture to the exclusion of other perennial plants. It should not be used as an indicator of change in range condition since it volunteers freely after rain and rapidly disappears from a community under drought conditions. However, it is a useful guide to the current level of use.

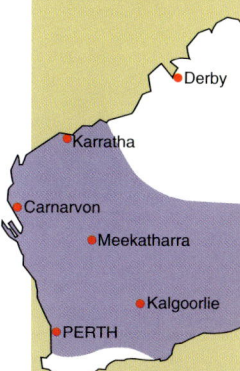

Flowers and mature fruit, about half actual size

353

Wild tomato   Round leaved solanum

Solanum orbiculatum

Family: Solanaceae

GENERAL DESCRIPTION: Wild tomato is a compact, round shrub up to 1 m high. It has large, strong spines up to 1 cm long on the stems. They are either straight or slightly curved, but unlike other *Solanum* species, they do not occur on the leaves or fruits. The stems and leaves have a sparse covering of hairs, which give the plant a dusty grey or rust-coloured appearance. The leaves are oval, 1.5–6 cm long and 1.5–3.5 cm wide and are borne on short stalks. The plant is never densely covered with leaves. The flowers are about 2.5 cm wide and consist of five separate, narrow, purple petals with five large, yellow stamens united to form a central cone. The flowers mature to form globular, multi-seeded berries, about 1.5 cm in diameter. These turn whitish yellow on maturity.

Wild tomato, about 70 cm high

HABITAT: Wild tomato is common in many environments in the Nullarbor and Goldfields. In the Murchison and Gascoyne it is restricted to sandplains, wanderrie banks and mulga grove communities on loams and sandy clay loams.

FORAGE VALUE: Many *Solanum* species contain alkaloids and wild tomato must, therefore, be suspected of toxicity. However, this is not a problem since the plant appears not to be grazed.

INDICATOR VALUE: When it dominates a shrubland to the exclusion of other species, the vegetation may be considered to be severely overgrazed. In these circumstances it is an indicator of poor range condition.

Flowers, immature fruit and spines, about actual size

Mulga broom bush   Stick-plant bush

Spartothamnella teucriiflora

Family: Chloanthaceae

GENERAL DESCRIPTION: Mulga broom bush is an upright, rigidly branched, multi-stemmed shrub that rarely exceeds 90 cm in height. Individual stems are cane-like and may be unbranched where growth is uninterrupted. The leaves are very small, 2–10 mm long, and are spaced at 2–4 cm intervals along the stems. They taper towards their ends and have rolled margins. The flowers are white and small and have five petals. They are succeeded by small, black, berry-like fruits which are green or red when immature.

HABITAT: Mulga broom bush grows on a wide range of sandy soils and sandy clay loams with mulga, tall saltbush and Wilcox bush.

Mulga broom bush, about 50 cm high

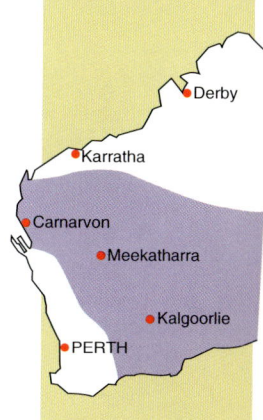

FORAGE VALUE: Under drought stress, the leaves are mostly shed and the branches die back towards the plant base. It is from this appearance that the plant derives its common name. When green, it is lightly grazed, but its overall contribution as a pasture species is low.

INDICATOR VALUE: Mulga broom bush declines in numbers under heavy and prolonged grazing pressure. Re-establishment is adversely affected by soil disturbance. An increase in numbers of this species indicates an improvement in range condition.

Leafless stems, about one and a half times actual size

357

Pebble bush

Stylobasium spathulatum

Family: Surianaceae

GENERAL DESCRIPTION: Pebble bush is an erect shrub growing up to 2.5 m. It has greyish green foliage and round, pebble-like fruit. The leaves are up to 3.5 cm long, 1 cm wide and are held on stalks up to 1 cm long. They are indented at their tips. The flowers are small and yellow with ten bright, yellowish brown stamens hanging out of the cup-like petals. The seeds are held in a hard, round, stony fruit up to 5 mm in diameter.

Pebble bush, about 1.8 m high

HABITAT: Pebble bush is found on a wide range of habitats from floodplains to sandplains. It grows on an equally wide range of soil types.

FORAGE VALUE: It is variable in palatability, being ignored in some situations and relished by stock in others. Crude protein levels of 16 per cent have been recorded.

INDICATOR VALUE: It tends to increase under moderate grazing pressure, but under heavy grazing it declines and may be absent from pastures in poor condition.

NOTE: It is one of the first species to colonise burnt sandplains in the Shark Bay area.

Immature fruit, about actual size

Fish poison   Corky bark caltrop, Skeleton bark

Tribulus platypterus

Family: Zygophyllaceae

GENERAL DESCRIPTION: Fish poison is a spreading shrub growing to 1 m. It has a prominent, white, corky bark and is never profusely leafed. The leaves consist of four pairs of dark green, hairy leaflets which are each up to 1 cm long and 5 mm broad. The flowers are a bright yellow and have five petals. They are 2 cm in diameter and have a sweet smell. The seeds are held in a large, papery, five-segmented fruit that is 2 cm in diameter and white in colour on maturity.

Fish poison, about 60 cm high

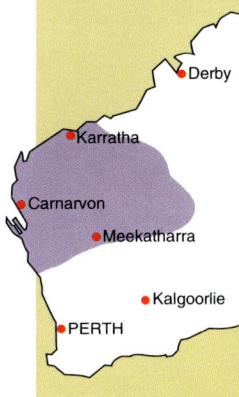

HABITAT: Fish poison is never common. It grows on shallow soils on stony hills.

FORAGE VALUE: It is readily eaten by sheep and cattle and, during periods of forage shortages, is eaten down to the large woody branches.

INDICATOR VALUE: It is a decreaser species. As it is sparse, it cannot be relied upon to indicate condition.

NOTE: The bark from this species is reputed to have been used by the Aborigines to stun fish.

Leaves and distinctive corky bark on main stem, about three-quarters actual size

Prominently winged fruit, about actual size

Shrubby twinleaf

Zygophyllum aurantiacum

Family: Zygophyllaceae

GENERAL DESCRIPTION: Shrubby twinleaf is a perennial, scrambling shrub growing to 1 m. The stems are slender and rigid, but brittle. The leaves are succulent and sparsely distributed along the stems. They are deeply lobed and consist of two leaflets supported by a stout stem of equal size. They are normally a light green tending to yellow-green in colour. The flowers are yellow and have four petals about 1 cm long. The fruit hangs pendulously from the stem. It is a capsule with four prominent, membranous wings and is about 2 cm across. The fruit splits at maturity into four triangular pieces, each containing a seed.

Shrubby twinleaf, about 40 cm high

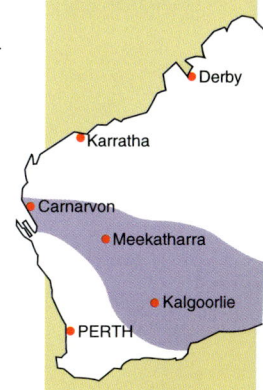

HABITAT: Shrubby twinleaf is restricted to scattered populations in semi-saline, lake-frontage communities including kopi (gypsum) dunes.

FORAGE VALUE: It is eaten by stock during periods of forage shortage. The foliage, however, does not persist for long during dry periods.

INDICATOR VALUE: It is a decreaser species, but because of its sporadic distribution it is not a reliable indicator of range condition.

Leaves and seeds, about half actual size

SECTION IV

Trees and shrubs
more than 3 m high

Acacia group

Members of the genus *Acacia* vary from 10 m high trees to 1 m high shrubs. In the arid interior, this group dominates most parts of the landscape, but is noticeably absent from highly saline habitats.

Acacia leaves are technically not leaves at all, but modified leaf stems. In Australia, the acacias have responded to the dry climate by shedding their leaflets and have modified their leaf stems to take the place of their leaves. These are called phyllodes. In this book we have called acacia phyllodes leaves. When an acacia seed germinates, the first two leaves have leaflets but these are quickly shed and all further leaves are modified leaf stems.

The group is characterised by numerous very small, yellow or creamy white flowers united into cylindrical or globular heads. The flowers form pods. Each acacia species is generally characterised by the shape of the flower heads, shape of the pods, type of bark, type of leaves and the shape of the bush.

There are about 1000 species of acacia in Australia, of which about 600 are found in Western Australia. The most common 25 species are described in this book.

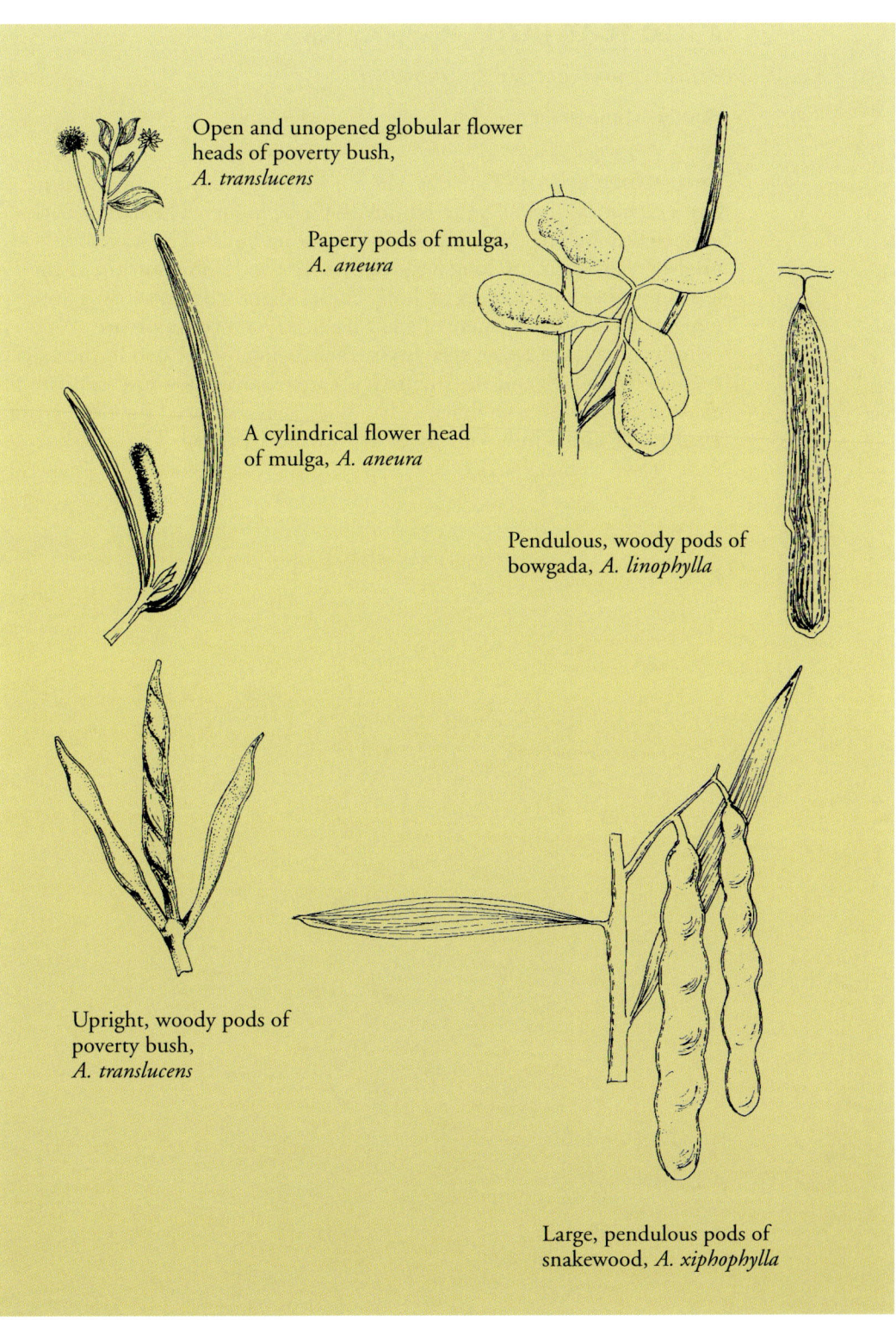

Flower heads and pods of different acacias

ACACIA GROUP

Fine leaf jam   Raspberry jam

Acacia acuminata subsp. *burkittii*

Family: Mimosaceae

GENERAL DESCRIPTION: Fine leaf jam is a small tree which grows to 4 m but which may, in well-watered situations, be higher. The bark is dark grey, tightly fissured and extends to the young limbs. The leaves are bright green, 8–12 cm long and about 2 mm wide. The soft tip of the leaf is pointed and characteristically curves slightly downwards. The margins have a dense fringe of short hairs. The flower heads are cylindrical and are borne very freely, there being one or two in the axils between the leaves and the stems. They are lemon yellow and consist of dense masses of minute flowers. The pods which succeed the flowers are 10 cm long and 5 mm wide, light brown and flattened. The seeds are dark brown to black.

Fine leaf jam, about 3–5 m high

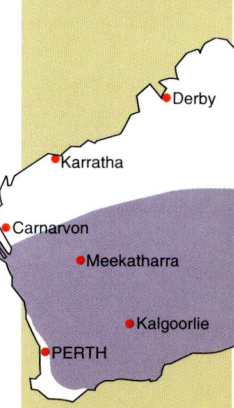

HABITAT: Fine leaf jam is usually found on rocky surfaces, in pockets of loamy soils on stony uplands and along some creeklines. It grows with Cue grevillea and limestone wattle and appears to favour neutral to acid soils. It is not as common in the shrublands as it is in the wheatbelt of the South-West Land Division.

FORAGE VALUE: Fine leaf jam does not have any forage value as most plants are out of reach of stock.

INDICATOR VALUE: Fine leaf jam has no indicator value in the assessment of range condition.

NOTE: Fine leaf jam is sometimes called raspberry jam, from the strong raspberry jam odour it emits when fresh wood is cut.

Leaves with characteristic point, about half actual size

ACACIA GROUP

Sandridge gidgee   Gidgee

Acacia anastema

Family: Mimosaceae

GENERAL DESCRIPTION: Sandridge gidgee is a tree growing to 7 m. It may sometimes be branched near ground level, producing two main trunks, though it more commonly has a single trunk. It has straight limbs producing a flat-topped cone of foliage. The bark is grey and rough on the trunk, but smooth on the branches. The leaves are up to 20 cm long and 5 cm wide, and are sickle-shaped. The flower heads are cylindrical spikes of many minute, bright yellow flowers and are 2–4 cm long. The seeds are carried in pods up to 14 cm long and 2–3 cm wide, slightly curved and very slightly

Sandridge gidgee, about 6 m high

constricted between the seeds. The seeds are cylindrical, greyish brown, and up to 5 mm long and 3 mm wide. They are held lengthwise in the pods.

HABITAT: It occurs on sand dunes and sandplains and grows with wanyu, sand-dune grevillea and limestone wattle.

FORAGE VALUE: It is unlikely to contribute to diet, since the canopy is out of the reach of stock. This species is cut for fence posts and prized for yards locally because of its straight limbs and resistance to white ants.

INDICATOR VALUE: It has no known indicator value.

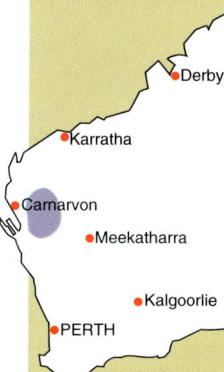

Flower heads, about three quarters actual size

ACACIA GROUP

Mulga

Acacia aneura

Family: Mimosaceae

GENERAL DESCRIPTION: Mulga is a shrub or small tree, ranging in height from 2 to 8 m. It is most variable in habit and shape. It may have a single stout trunk bearing a canopy far off the ground, or it may be multi-stemmed and almost 'mallee'-like in form. The leaves are particularly variable. They may be long, green and thread-like, or broad, elliptical and grey. All have a light waxy covering and appear to be veined in a parallel fashion. The flower head is cylindrical, being about 2 cm long and 4 mm across and composed of many tightly compressed flowers. The seeds are held in flat, papery pods, which do not have prominent veins.

HABITAT: Mulga occurs in most habitats, though it avoids heavy clay soils.

Shrub-like mulga, about 3.5 m high

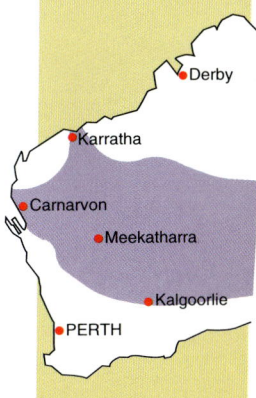

FORAGE VALUE: Fresh mulga leaves are a useful forage for stock and contain from 6 to 12 per cent crude protein. As much of the canopy is out of reach, the majority of the fresh material is unavailable. Mulga gives stability to many plant communities. It is historically an important source of timber for fences, yards and buildings.

INDICATOR VALUE: Mulga is not a good indicator of range condition. It is very resistant to grazing and is one of the last plants to disappear from overgrazed communities. Some thickening of mulga stands has been observed in the Murchison as young seedlings, which volunteered in the good rainfall years of the 1960s, have become established. This increase in the mulga cover will reduce the forage-producing capability of the land, since the new trees will deny space to more palatable low shrub species.

Flowers and foliage of broad leaved mulga, about actual size

Mulga, about 7 m high. Note the single trunk

Immature pods and foliage of needle leaved mulga, about actual size

Immature pods and foliage of narrow leaved mulga, about actual size

ACACIA GROUP

Umbrella mulga False bowgada

Acacia brachystachya

Family: Mimosaceae

GENERAL DESCRIPTION: Umbrella mulga is a tall, bushy shrub with twisted and spreading stems, which develop into a dense, wide crown often broader than the shrub is high. It may grow to 5 m, but it is commonly much less. The bark is dark grey and deeply fissured. The leaves are variable in size and length. They are up to 15 cm long, mostly 1–2 mm wide and narrow, though occasionally they may be sub-circular in cross-section. The yellow wattle-like flowers are gathered in cylindrical bundles and are up to 2 cm long. They are succeeded by long, straight pods up to 8 cm long and 8 mm across. They may be constricted between the seeds and have long striations on their sides.

Umbrella mulga, about 3 m high

They are convex on each side when ripe. The pods may be used to distinguish this species from true bowgada, in which the pods are truly cylindrical in cross-section.

HABITAT: It is found growing in mulga communities on deep soils as well as on rocky hills.

FORAGE VALUE: It is not readily eaten and it is not as palatable as mulga. The leaves contain up to 13 per cent crude protein with a digestibility of 37 per cent.

INDICATOR VALUE: It has no indicator value as it only occurs as scattered trees.

Leaves and pods, about half actual size

ACACIA GROUP

Black mulga

Acacia citrinoviridis

Family: Mimosaceae

GENERAL DESCRIPTION: Black mulga is a tree growing to about 8 m. It rarely has more than one trunk. The bark is dark grey and well fissured even on the younger branches. The leaves are a grey-olive colour and are up to 12 cm long and about 1 cm wide. They are broadly elliptical in shape, tending to be slightly curved in some instances. The flowers are small cylinders of tightly packed florets, which are succeeded by pods about 8 cm long and up to 1 cm across. The pods have a lemon-green, felty covering from which the botanical name is derived. The seeds are dark brown to black. The tree is similar to mulga at a distance but it may be distinguished by the dark bark, and the pods when in season.

Black mulga, about 6 m high

HABITAT: Black mulga is the common acacia growing on the margins of creeklines and rivers north of the Murchison. It is associated with river gums, mulga and curara. It is not found away from well-watered situations.

FORAGE VALUE: It does not have any forage value as most leaves are out of reach of stock.

INDICATOR VALUE: Black mulga has no indicator value.

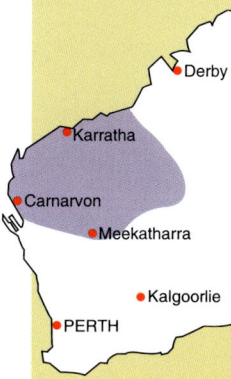

Leaves and pods, about three-quarters actual size

ACACIA GROUP

Sugar brother Spinifex wattle

Acacia coolgardiensis

Family: Mimosaceae

GENERAL DESCRIPTION: Sugar brother is a tall shrub with multiple stems growing to 3 m. It has a spreading habit and grows on spinifex sandplains. The green leaves are up to 10 cm long and 3 mm wide, flat and slightly curved with a blunt hook at the tip. They are held vertically. The flower heads consist of a cylindrical mass of minute yellow flowers, up to 2 cm long and 5 mm wide. They are borne on a short stalk. The seeds are held in papery pods up to 3 mm in diameter.

Sugar brother, about 3 m high

HABITAT: Sugar brother grows in association with hard spinifex and *Acacia jamesiana* on sandplains with deep, red, loamy sands.

FORAGE VALUE: It has no known forage value.

INDICATOR VALUE: This species has no known indicator value.

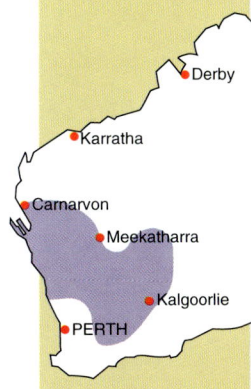

Immature pods, about actual size

ACACIA GROUP

River jam   Wiry wattle

Acacia coriacea

Family: Mimosaceae

GENERAL DESCRIPTION: River jam is a small tree that ranges in height from 3 to 8 m. It generally has one or two main stems. The bark is dark grey and deeply furrowed. The foliage droops from the branches, the leaves being 20–30 cm long, 1–2 mm broad, thick and leathery with many fine veins. They are a mid-green in colour, though the young leaves are slightly yellow. The globular flower heads are yellow, about 5 mm in diameter and are

River jam, about 6 m high

composed of many tightly compressed flowers. The distinctive pods are 15–23 cm long and 6–12 mm wide. They are twisted and constricted between each seed. The large seeds have a distinctive red appendage or aril.

HABITAT: River jam occurs as a spreading, low tree behind coastal dunes. It is also found as a tall tree on the banks of large rivers and as a low tree on hard spinifex sandplains.

FORAGE VALUE: The tree is eaten by cattle during low rainfall periods, especially on the river frontages. It has moderate nutritive value.

INDICATOR VALUE: This species has no known indicator value.

Leaves and pods, about half actual size

ACACIA GROUP

Hop mulga

Acacia craspedocarpa

Family: Mimosaceae

GENERAL DESCRIPTION: Hop mulga is a shrubby tree which can reach 3 m. It is usually rather rounded in its general form. The leaves are bluish green, flattened, rather broader at the end and overall are usually 2 cm long and 1 cm wide. The flower heads are cylindrical, being up to 2 cm long and 5 mm across. The woody pods are broad, flattened, much thickened by surface ribbing and resemble the pods of commercial hops.

HABITAT: Hop mulga is common in watercourses and occurs as a fringing community along the margins of the more defined streamlines.

Hop mulga, about 2 m high

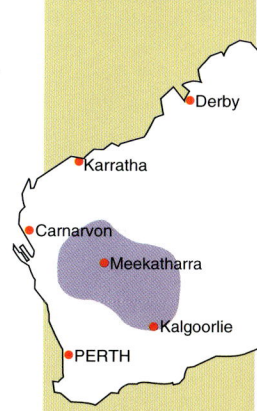

FORAGE VALUE: The plant is rarely eaten and has little forage value. It can become very dense and excludes other plants when it grows in thickets. Pastures dominated by hop mulga provide very little feed even in good seasons.

INDICATOR VALUE: Hop mulga has no known indicator value for the assessment of range condition.

NOTE: Mulga and hop mulga are often difficult to tell apart. However, they can be separated by the venation on the leaves. Hop mulga has branching veins, whilst mulga has parallel veins.

Immature pods, about actual size

ACACIA GROUP

Wait-a-while Bohemia

Acacia cuspidifolia

Family: Mimosaceae

GENERAL DESCRIPTION: Wait-a-while is a robust tree which grows to 7 m. It has a gnarled appearance and a rough, dark brown bark. The leaves are 3–7 cm long, 2–5 mm wide and are straight to slightly curved. They are flat and olive green when mature, but are light green when young. The leaf tips come to an abrupt blunt end and terminate in a spine 2 mm long. Strong hooked spines up to 5 mm long occur in the axils of leaves and stems. It has pale yellow, globular flower heads 5 mm in diameter, a pair being produced from each leaf axil. Flattish, oblong, light brown pods 5–9 cm long and 1–2 cm wide are produced and contain hard, flattened, black seeds.

Wait-a-while, about 3 m high

HABITAT: Wait-a-while grows on floodplains and undulating country on alkaline clays or loams or on texture contrast soils. It also occurs on saline sluggish drainages and in crabholes in stony uplands. It grows with Gascoyne bluebush, river saltbush, golden bluebush, snakewood, prickly acacia and ruby saltbush.

FORAGE VALUE: The foliage persists during droughts and is eaten during periods of feed shortage. It is especially favoured by cattle. The crude protein content varies from 9.5 to 11 per cent, and the digestibility from 26 to 33 per cent. The hooked spines tear at the wool of sheep grazing the trees and in droughts they are bedecked with wool. In some circumstances weak sheep become trapped by the spines in the foliage and, being unable to retreat, perish, hence the common name.

INDICATOR VALUE: When edible shrub populations growing with this species are suppressed by overgrazing, populations of wait-a-while increase. A large build-up of this species, without edible shrubs, indicates poor range condition.

Mature pod, about half actual size

Curved spines on stems, about half actual size

ACACIA GROUP

Creekline miniritchie   Red mulga

Acacia cyperophylla

Family: Mimosaceae

GENERAL DESCRIPTION: Creekline miniritchie is a tree with a single straight trunk, which grows to 7 m. It may also branch near the ground to produce two or three main trunks. The trunks and branches are covered in bright orange to red, curling, miniritchie-type bark with frequent bald spots. The leaves are green, about 10–15 cm long and are rigid and curved. They are almost round in cross-section, about 2 mm in diameter and often end in a point. The flower heads, borne after rain, consist of cylindrical spikes of minute flowers and are up to 2 cm long. The pods are broad and flattish, up to 8 cm long and 7 mm wide. The seeds lie longitudinally in the woody pod and are about 8 mm long.

Creekline miniritchie, about 6 m high

HABITAT: Creekline miniritchie grows on the banks of the larger sandy creeks. It grows with curara, and with an under-storey of weeping grass, flood grass and buffel grass.

FORAGE VALUE: It has no forage value since the leaves are out of the reach of grazing animals.

INDICATOR VALUE: It has no indicator value.

NOTE: The wood of creekline miniritchie is very hard when dry and is excellent for campfires.

Red curly bark on trunk, about 15 cm in diameter

389

ACACIA GROUP

Miniritchie

Acacia grasbyi

Family: Mimosaceae

GENERAL DESCRIPTION: Miniritchie is a shrubby tree which grows to 4 m. It has twisted stems and a reddish brown bark which peels continually in small curly flakes. It has a flat-topped appearance. The leaves, which are held vertically, are up to 9 cm long and 1 mm wide, round in cross-section and have bluntly pointed tips. The flower heads consist of cylindrical spikes of many minute, yellow flowers, are up to 3 cm long and 5 mm diameter and are held on stalks about 2 cm long. The seeds are large, up to 6 mm long, and are black. They are held in roundish, brown pods up to 11 cm long with tight constrictions between the seed compartments.

Miniritchie, about 2.5 m high

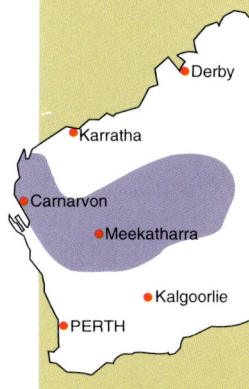

HABITAT: Miniritchie is mainly found on hardpan plains and rocky hills with shallow, skeletal, loamy soils. It grows with mulga and other shrubs and trees common in wind grass communities.

FORAGE VALUE: It is palatable to stock and contains up to 12 per cent crude protein.

INDICATOR VALUE: Under excessive grazing, populations of miniritchie will decrease but it is very tolerant of harsh conditions and consequently cannot be used as a sensitive indicator of range condition.

NOTE: Miniritchie should be distinguished from creekline miniritchie, which is restricted to creeklines and is a tall, open tree with rough, dark orange bark.

Dark reddish brown, curly bark on trunk, about 15 cm in diameter

Mature pods, about one-third actual size

ACACIA GROUP

Tan wattle

Acacia hemiteles

Family: Mimosaceae

GENERAL DESCRIPTION: Tan wattle is a bushy wattle, which grows to about 3 m and is often broader than it is high. The abundant stems and branches are smooth-barked and a mid-brown or grey in colour. The rigid leaves are grey-green and are narrow, being about 3–4 mm wide and 7–8 cm long. The tips are strongly pointed. Leaves are arranged alternately along the stems and are borne freely. The flowers are held in yellow, globular heads arising in the axil between stem and leaf. The brown pods are about 1 cm across and up to 10 cm long. They are laterally constricted between the seeds.

HABITAT: Tan wattle is a volunteer species colonising disturbed areas with vigour. It is an early coloniser after fire and is common around the mining areas where the ground is disturbed. It grows on sandy loams and sands with bowgada, mulga and shrubs.

FORAGE VALUE: Stock do not appear to eat it.

Tan wattle, about 3.5 m across

INDICATOR VALUE: It has no indicator value.

REHABILITATION: Tan wattle is a useful species to include in seeding mixes applied to mine dumps which are not highly saline. It will add substantially to the structure of the stand being established, being taller than other plants usually included in rehabilitation work. The seed may need pre-treatment with hot water or by scarification before planting. About 1 kg/ha should be included in seed mixes. Seed may be collected locally as it seeds prolifically.

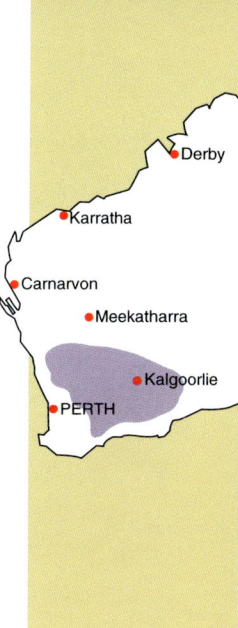

Leaves and pods, about half actual size

ACACIA GROUP

Kanji bush

Acacia inaequilatera

Family: Mimosaceae

GENERAL DESCRIPTION: Kanji bush is a gnarled tree with one or two main stems that have corky bark, rather similar to that of corkwood. It is 2–5 m high and has spiny foliage and stems. The leaves are bluish grey and are 2.5–7 cm long, 1.5–3.5 cm wide, terminating in a short spine. The midrib is curved, giving the leaves a lop-sided or angular appearance, leading to the botanical name. A pair of recurved spines or hooks is located at the base of each leaf. The globular flower heads consist of numerous minute florets.

Kanji bush, about 4 m high

They are purplish yellow in colour and are up to 5 mm in diameter. The seed pods are flat-sided, curved and are up to 10 cm long and 1 cm wide.

HABITAT: Kanji bush is common on sandplains that support soft spinifex. It is very fire tolerant and volunteers rapidly after a burn. It is comparatively short-lived, lasting less than ten years. It also occurs on basalt hills.

FORAGE VALUE: Sheep eat kanji bush during periods of forage shortage, but cattle rarely eat it.

INDICATOR VALUE: It has no indicator value as the spinifex sandplains are frequently burnt.

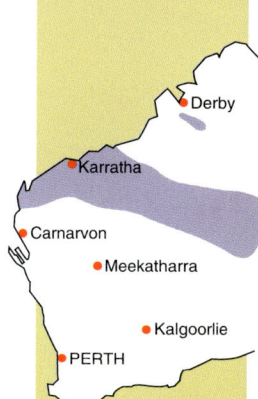

Leaves showing short spines and their angular appearance, about one-quarter actual size

ACACIA GROUP

Wanderrie wattle   Witchetty bush, Granite wattle

Acacia kempeana

Family: Mimosaceae

GENERAL DESCRIPTION: Wanderrie wattle is a spreading shrub which grows to 4 m. It has many stems which branch at the base to form a cone-shaped bush. The noticeably green leaves are broad and flat, up to 9 cm long and 1.5 cm wide, moderately curved and generally pointing vertically. The flower heads form in September/October and consist of many minute, yellow flowers, clustered to form a cylindrical yellow spike 1–2 cm long. The black seeds are held in flat, papery, brown pods up to 7 cm long and 1.5 cm wide. Constrictions across the pods separate the seeds.

Wanderrie wattle, about 2 m high

HABITAT: Wanderrie wattle is found on deeper, sandy soils such as wanderrie banks, along some creeklines and adjacent to granites in the Goldfields. It grows with wanderrie grasses, Wilcox bush, various cassias and mulga, and is never found in dense populations.

FORAGE VALUE: Stock graze on wanderrie wattle during droughts as a 'last resort' fodder. It has a crude protein content of about 8 per cent.

INDICATOR VALUE: Populations decline under severe overgrazing, but it is not a sensitive indicator of range condition.

Immature pods, about actual size

ACACIA GROUP

Umbrella wattle

Acacia ligulata

Family: Mimosaceae

GENERAL DESCRIPTION: Umbrella wattle is a fast-growing and bushy, spreading shrub that rarely exceeds 4 m in height. The leaves vary greatly, from 4 to 10 cm in length and 5 to 10 mm in width. They are linear or linear-lanceolate in shape. They are generally thick in cross-section, wrinkling when dry. The flowers arise from the axils between stem and leaf, either singly or in small racemes. They are globular, an orange-yellow in colour and consist of many minute flowers. The pods are woody and constricted between the seeds and measure up to 12 cm long and 1 cm wide.

Umbrella wattle, about 3 m high

HABITAT: It grows on deep, sandy soils and on sand dunes in association with bowgada, mulga, Wilcox bush and cotton bush.

FORAGE VALUE: Umbrella wattle is not highly palatable. Sheep commonly skirt shrubs close to water, up to a height of 1.3 m.

INDICATOR VALUE: Umbrella wattle has a low indicator value. It can be expected to increase rapidly where historical overgrazing has occurred and pastures are in a recovery phase.

Leaves and flowers, about two-thirds actual size

ACACIA GROUP

Bowgada   Wanyu

Acacia linophylla

Family: Mimosaceae

GENERAL DESCRIPTION: Bowgada is a spreading shrub or low tree which grows to 4 m. It produces multiple stems from the base and these form a shrub that is often wider than it is high. The bark is grey and smooth. The leaves are greyish green in colour, almost round in cross-section and bluntly pointed. They are held almost vertically on the bush and are up to 13 cm long and 2 mm across. The flower heads are borne in September and October and consist of many minute flowers packed together to form cylindrical yellow spikes up to 2 cm long. The 5 mm long seeds are black, and are held in cylindrical pods up to 10 cm long and 1 cm wide. The pods have deep longitudinal fissures along their entire length. The woody pods are filled with pith, which serves to separate the seeds. The pods usually fall entire and do not open at maturity.

Bowgada, about 2 m high

HABITAT: Bowgada grows on wanderrie banks and sandplains, where the soils are loose, deep, red sands or loams. It grows with mulga, wanderrie grasses, eucalypts and grevilleas.

FORAGE VALUE: The foliage is rarely eaten, but the seed and pods contain over 20 per cent crude protein and are sought after by sheep. They are, therefore, an important forage in some seasons. Insect damage to pods is always high.

INDICATOR VALUE: This species has no indicator value in the assessment of range condition.

NOTE: *A. linophylla* is very similar to *A. ramulosa*. They differ only in leaf shape. *A. linophylla* has leaves which are round in cross-section whilst *A. ramulosa* has flat leaves, often only 5 mm wide. *A. ramulosa* is called wanyu in the western Gascoyne.

Mature pod, about three-quarters actual size

ACACIA GROUP

Sandplain wattle Murray's wattle, Fire wattle

Acacia murrayana

Family: Mimosaceae

GENERAL DESCRIPTION: Sandplain wattle is a bushy, fast-growing, tall shrub or shrubby tree, reaching 8 m. It has rough, dark grey bark on the larger limbs, while the younger branches have a conspicuous white, powdery covering that can be rubbed off easily. The leaves are 5–12 cm long and 2–7 mm wide, straight-sided and rounded at the tip. They are grey or pale green. The bright yellow flowers are borne in globular heads up to 8 mm in diameter. Each flower head consists of about 40 minute flowers and they appear in September and October. The pods are 5–8 cm long and up to 1 cm wide. They are flattish and papery, often covered with a mauve bloom, and contain hard, greyish black seeds.

Sandplain wattle, about 4 m high

HABITAT: It occurs on sand ridges and disturbed areas which have deep, red, acid sands. It grows with bowgada, mulga, limestone wattle, desert poplar and oat-eared spinifex. It can also form thickets as it suckers freely.

FORAGE VALUE: The leaves are rarely eaten, though the pods are sought after by stock.

INDICATOR VALUE: Its presence in large numbers indicates site disturbance by fires, overstocking or road works.

NOTE: It has a very distinctive 'rank' smell when the leaves are crushed.

Leaves and pods, about half actual size

Flower heads, about half actual size

403

ACACIA GROUP

Western myall

Acacia papyrocarpa

Family: Mimosaceae

GENERAL DESCRIPTION: Western myall is a handsome tree which grows to 7 m. It has a dark grey, rough bark and a dense crown. The leaves are straight, greyish green and flattish, 4–12 cm long and 1–2 mm wide. The young foliage has a silvery appearance. The flower heads consist of numerous small flowers compressed into yellow balls about 5 mm in diameter. The seeds are held in long, thin, flat pods about 11 cm long and 4–10 mm wide.

HABITAT: Western myall grows on limestone plains and on highly alkaline loams or clays with pearl bluebush, broom bush and bladder saltbush.

Western myall, about 5 m high

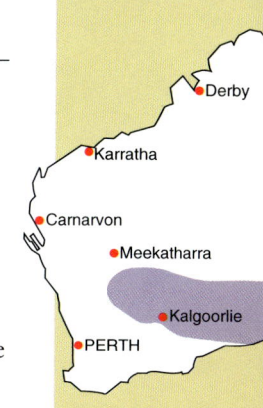

FORAGE VALUE: It is grazed by stock during times of forage shortage. Stock and rabbits favour the young seedlings and selectively graze them. It is rare to find young plants in stocked paddocks. Studies on this species in South Australia have shown that myalls do not survive under grazing. If the young seedlings are not given an opportunity to develop and establish, this attractive tree will disappear as the older members of the population die.

INDICATOR VALUE: It is very long-lived and cannot be used as an indicator species for range condition assessment.

NOTE: A very similar species, northern myall (*Acacia sibilans*), is found in the Murchison where it grows only on limestone.

ACACIA GROUP

Gidgee   Black gidgee

Acacia pruinocarpa

Family: Mimosaceae

GENERAL DESCRIPTION: Gidgee is a tree which grows to 12 m. It has a rough, dark brown bark on gnarled trunks, and can occur in quite dense stands when young. Large trees, reaching to 12 m with a girth of 2 m, are common in broad watercourses and drainage foci. The leaves have blunt tips and are greyish green, up to 17 cm long and 2 cm wide, and gently curved. It flowers from October to December and the seed matures in January to February. The flower heads are globular and yellow and consist of numerous minute flowers. The seeds are produced in papery, pale brown pods up to 12 cm long and 2 cm wide. The pods are flattened and not constricted between the seeds and are covered with a whitish bloom.

Gidgee, about 8 m high

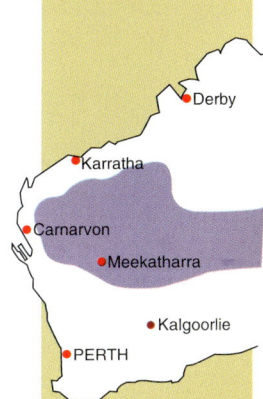

HABITAT: Gidgee occurs on rocky hills and on hard mulga plains with shallow loamy soils over rock or hardpan. It also grows on the better-watered sandplain margins and in broad watercourses and drainage foci. It grows with mulga, beefwood, curara, cotton bush, tall saltbush and oat-eared spinifex, depending on its location.

FORAGE VALUE: It is readily eaten by sheep and cattle during periods of feed shortage. It is extremely drought tolerant and the foliage persists well, making this species a good drought reserve. Most of the canopy of older trees, however, is out of the reach of stock. It contains about 13 per cent crude protein and is high in total minerals.

INDICATOR VALUE: It has no indicator value for range condition assessment.

Immature pods, half actual size

ACACIA GROUP

Granite wattle   Spreading wattle

Acacia quadrimarginea

Family: Mimosaceae

GENERAL DESCRIPTION: Granite wattle is a low tree which grows to 6 m, and is frequently wider than it is high. The bark is mid-grey in colour and not prominently fissured. The leaves, which are curved to elliptical in shape, are a bright mid-green with a faint reddish margin. They hang down on slender stems and are up to 10 cm long. The yellow flowers are held in tight cylindrical spikes and are produced in spring and summer. They are followed by thick, four-sided pods which are up to 15 cm long.

Granite wattle, about 6 m across

HABITAT: Granite wattle is found on shallow soils within granite complexes. It is also found on sandy loams and sandy clay loams with mulga and is frequent along sharply incised creeklines in hills and rocky uplands. It commonly grows with poverty bushes, flannel bush and cotton bush.

FORAGE VALUE: It has limited forage value as it does not appear to be attractive to stock.

INDICATOR VALUE: It has no indicator value.

Leaves and seed pods, about half actual size

ACACIA GROUP

Limestone wattle Silver bark wattle

Acacia sclerosperma

Family: Mimosaceae

GENERAL DESCRIPTION: Limestone wattle is a spreading, dense, tall shrub or shrubby tree, up to 4 m high and 6 m in diameter. The bright green leaves are up to 7 cm long and 2 mm wide, and are oval in cross-section. The flower heads, which appear opportunistically after rain, consist of round, yellow, globular heads 5 mm in diameter and are composed of many minute flowers. The large black seeds are up to 1 cm long and are held in pods up to 12 cm long with marked constrictions between the seeds.

HABITAT: Limestone wattle grows on floodplains, calcrete platforms, limestone valleys, sandplains, creeklines, river banks and other areas with alkaline loams. It grows with river saltbush and a wide variety of species characteristic of major flow lines and calcareous areas, and also with spinifex on sandplains that overlie calcrete.

Limestone wattle, about 2 m high

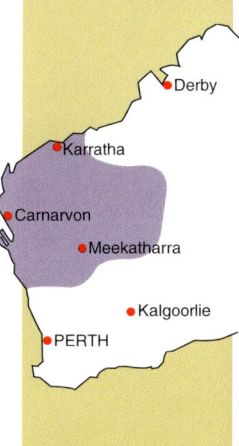

FORAGE VALUE: It is not highly palatable and is only eaten as a forage of last resort during dry periods. The crude protein content is about 8 per cent and digestibility, 45 per cent.

INDICATOR VALUE: It is sometimes an increaser species if palatable perennial grass and shrub species are overgrazed on floodplain and limestone areas. Generally, however, it has no indicator value. In heavily overused vegetation, limestone wattle can be pruned to sheep grazing height; such vegetation is usually in poor range condition.

Almost mature pods, about three-quarters actual size

ACACIA GROUP

Spreading gidgee

Acacia subtesserogona

Family: Mimosaceae

GENERAL DESCRIPTION: Spreading gidgee is an upright, fast-growing tree that commonly grows to 7 m. The leaves are sickle-shaped, and about 11 cm in length with a maximum mid-section width of about 7 mm. The pods are 6–12 cm in length by about 8 mm wide, with a noticeable groove along each edge and a reticulated venation pattern.

HABITAT: It grows on loamy and sandy loam soils. On sand dunes, spreading gidgee is confined to the lower slopes and swales.

Spreading gidgee, about 6 m high

FORAGE VALUE: Spreading gidgee has negligible forage value and is rarely grazed, even when young. The upright habit that characterises the species places most foliage well above grazing height as trees mature.

INDICATOR VALUE: Spreading gidgee appears to be a long-term increaser species under heavy grazing. However, it is difficult to distinguish pastures naturally dominated by the species from those where the dominance can be attributed to grazing management.

Leaves and a pod, about three-quarters actual size

ACACIA GROUP

Curara Kurara

Acacia tetragonophylla

Family: Mimosaceae

GENERAL DESCRIPTION: Curara is a tall shrub or small tree, which can grow to 4 m. The bark on the young stems and branches is light grey and smooth. With advancing maturity the bark darkens and becomes deeply fissured. The leaves are sharply pointed, without hairs and usually are gathered into clusters. They are slender, needle-like, and up to 3 cm long and 1 mm wide. When young they are soft and pliable, but when mature they are rigid and very sharp. The flower heads are pale yellow and globular, borne either singly or in groups on stalks up to 3 cm long. The papery pods are very twisted, up to 8 cm long and about 5 mm wide. At maturity they break open and the seeds are shed.

Curara, about 3 m high

HABITAT: Curara is prominent on heavy loams along watercourses and on lighter soils in creeklines. It is also found on shallow soils, floodplains and on the skeletal soils of hills and rocky slopes.

FORAGE VALUE: The crude protein content ranges from 6 to 16 per cent. Curara is very drought resistant and is grazed, especially by goats, during periods of stress when other species have dropped their leaves. After grazing or leaf fall due to drought, new leaves appear on old branches. Curara always produces forage for stock as the lowest limbs do not remain bare, in contrast to other *Acacia* species.

INDICATOR VALUE: Curara has little indicator value since it persists under extremely harsh conditions.

Almost mature pods, about half actual size

ACACIA GROUP

Prickly acacia Bardi bush

Acacia victoriae

Family: Mimosaceae

GENERAL DESCRIPTION: Prickly acacia is a straggly, tall tree or shrubby tree, which grows to 6 m high and 10 m wide. It is also common as a low shrub up to 3 m high. It has short, flattened, curved, yellowish green leaves up to 3 cm long and 5 mm wide. There are a pair of straight spines up to 5 mm long at the base of most leaves. It flowers from August to October and seed matures from October to December. The flower heads are pale yellow and globular and made up of many small flowers. The seeds are almost round and slightly flattened on two sides. They are carried in golden brown papery pods, up to 9 cm long and 1.5 cm wide.

HABITAT: Prickly acacia grows on floodplains, on valley floors with saline soils and with spinifex growing on soils underlain by calcretes. It also occurs with snakewood in upland sluggish drainages on rocky slopes. It grows with river saltbush, sago bush, buffel grass, needlebush, fuchsia bush, tall saltbush and *Cassia* species.

Prickly acacia, about 3 m high

FORAGE VALUE: Animals readily eat the flowers and browse the foliage during dry periods. The leaves drop at the onset of severe moisture stress and once devoid of leaves the older branches rarely shoot again. The crude protein content is about 12 per cent and digestibility, 48 per cent.

INDICATOR VALUE: Vegetation dominated by this species south of the Pilbara is usually in poor range condition since prickly acacia inhabits sites generally occupied by bluebushes and saltbushes. It is a normal component of spinifex communities over calcretes in the Pilbara, but has no indicator value there. In some circumstances it is a useful species, as it has the ability to colonise bare, eroded alluvial plains and provide niches where other more edible species such as tall saltbush may establish.

REHABILITATION: Prickly acacia could be used in rehabilitation programs where soils are moderately saline. It is also capable of growing in alkaline situations. Seed should be given pre-treatment before sowing. Seeding rates of 500 g/ha are suggested.

Maturing pods, about one-third actual size

Flower heads, about half actual size

Spines on a stem, about two-thirds actual size

ACACIA GROUP

Snakewood

Acacia xiphophylla

Family: Mimosaceae

GENERAL DESCRIPTION: Snakewood is a spreading tree, up to 5 m high and 8 m wide. It has two or three twisted trunks with dark grey, rough bark. The bluish grey leaves are up to 8 cm long and 8 mm wide. They are flat or sometimes slightly curved and end in a point. Snakewood flowers opportunistically in response to rain and the flower heads consist of many minute yellow flowers tightly packed on a 3 cm long cylindrical spike. The pods form in clusters, each pod being up to 12 cm long and 8 mm wide, with constrictions between seeds. The seeds form horizontal bulges in the pods.

HABITAT: Snakewood occurs on saline and calcareous flats or hillslopes with clay or texture contrast soils. It grows with river saltbush, sago bush, and Gascoyne bluebush and bindiis. It is very common on the upland sluggish saline drainages of the Gascoyne and Ashburton catchments.

Snakewood, about 3.5 m high

FORAGE VALUE: It is only eaten as a last resort in times of severe forage shortage.

INDICATOR VALUE: It is usually a stable part of the vegetation, but may increase with heavy grazing. It has no indicator value itself, but its presence indicates those environments which should support valuable perennial shrub species such as bluebush and saltbush.

NOTE: *Acacia intorta* and *A. eremaea* are closely related species that appear similar to snakewood. *A. intorta* differs from *A. xiphophylla* in that it has long, round, needle-like leaves and is found only in the eastern Pilbara. *A. eremaea* differs very little from *A. xiphophylla* and is found in the Murchison.

Almost mature pods, about one-third actual size

Desert kurrajong

Brachychiton gregorii

Family: Sterculiaceae

GENERAL DESCRIPTION: Desert kurrajong is a single-trunked tree which grows up to 6 m high. The bark is finely fissured. The leaves are a shiny bright green and cause the tree to stand out in the drab mulga communities in which it is found. They are divided into three to five long, thinnish lobes up to 14 cm long and are deciduous. The flowers occur in small clusters on branchlets which spring from the axils between leaf and stem. They are small, yellowish, about 2 cm in length, roughly bell-shaped and appear in an opportunistic fashion after rain. The fruit is woody and shaped rather like a small boat. It is about 5 cm long, beaked at one end and contains 12 seeds.

Desert kurrajong, about 6 m high

HABITAT: Desert kurrajong can occur on sandy loams with mulga and other *Acacia* species. It also is common on the sides of some rocky ridges and low hills in the Goldfields.

FORAGE VALUE: Desert kurrajong is moderately palatable and nutritious.

INDICATOR VALUE: It has no indicator value.

NOTE: The hairs which pack the seed pods are highly irritating and should be avoided.

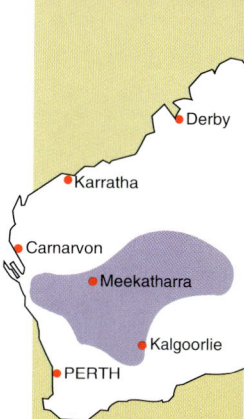

Leaves and fruit, about one-fifth actual size

Pine   White cypress pine

Callitris glaucophylla

Family: Cupressaceae

GENERAL DESCRIPTION: Pine is an upright tree with a single trunk, growing to 12 m. The leaves are small scales, a millimetre or two in size, growing in whorls on the branchlets. They are very aromatic when crushed. The flowering parts in pine are different from those of most of the plants described in this book. The flowers are incomplete and consist only of stamens in the male flower and naked ovules in the female. There are no other floral parts such as petals or sepals. The stamens are gathered into drooping spikes, while the ovules are found within whorls of scales, with

Pine, about 8 m high

several ovules attached to each scale. The seeds are held in a globular, finely wrinkled, woody cone about 2 cm in diameter. The seeds are released when the six segments of the cone break open.

HABITAT: Pine is found on sandy soils, particularly in the band of sandplain and dunes along the southern boundary of the shrublands.

FORAGE VALUE: As far as is known, this species is inedible to stock.

INDICATOR VALUE: It has no indicator value.

NOTE: This species is prized locally as a fencing timber and has been cut out in some areas. It is also fire-sensitive; large wild fires kill it, thus making its regeneration dependent on germination from seed. While this tree is not a true pine, it is called such because of its shape and colour and the smell of its leaves.

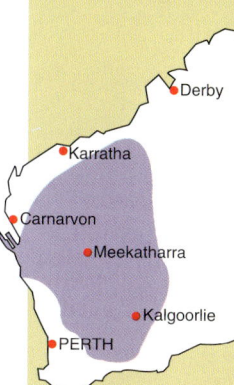

Leaves and cones, about half actual size

Wild lemon Wild currant, Native plum

Canthium latifolium

Family: Rubiaceae

GENERAL DESCRIPTION: Wild lemon is a small, often multi-stemmed tree which grows to 3 m. The bark is a mid-grey and, in most cases, smooth and not fissured. The leaves are a mid-green in colour and stand out prominently against the background of dull grey-green foliage of the mulga with which it is usually found. They are rather thick and stiff, strongly veined, rough to the touch and up to 10 cm long and 4 cm wide, being broadly similar to the leaves of the domestic lemon tree. The flowers, which emerge in autumn and

Wild lemon, about 4 m high

summer, are held in groups on short stems. They are white to cream in colour and up to 1 cm across, having five petals. They resemble the flowers of the lemon and are scented. The fruit is a black berry the size of a currant.

HABITAT: Wild lemon is closely associated with mulga and is found most commonly in mulga grove communities. It is found on deep loams and sandy clay loams but not on shallow, rocky soils.

FORAGE VALUE: There are reports of the plant being poisonous but as it is rarely eaten, its toxic properties are probably not significant.

INDICATOR VALUE: Wild lemon has no indicator value.

NOTE: Other species of *Canthium, C. lineare* and *C. attenuatum*, with narrow leaves, are also found in mulga groves. *C. lineare* is palatable to stock.

Leaves and fruit of C. lineare, *about one-quarter actual size*

Leaves and fruit of C. latifolium, *about one-tenth actual size*

Sheoak Swamp oak

Casuarina obesa

Family: Casuarinaceae

GENERAL DESCRIPTION: Sheoak is a tree up to 16 m high, with a dark grey-black bark covered in fine, 1 cm deep fissures. The 'leaves' consist of upright, joined branchlets, 2 mm in diameter and up to 25 cm long. There are many branchlets along each stem. There are separate male and female plants. The male flowers consist of spikes about 2 cm long, borne towards the ends of the branchlets. The female flowers are found in short, compact heads at the base

Sheoak, about 7 m high

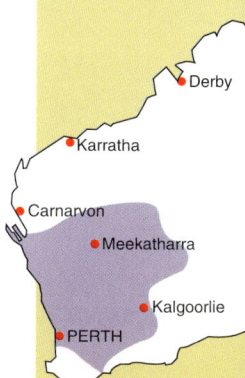

of the smaller branches. The seeds are carried in beaked structures massed into cylindrical nuts about 1.5 cm in diameter and 1 cm long. The winged seeds are released from the nut when the two prominent valves of the beaks open.

HABITAT: Sheoak grows on undulating alkaline clay plains, which generally support a chenopod shrubland of pearl bluebush and sago bush.

FORAGE VALUE: It is not known to be edible to stock.

INDICATOR VALUE: It has no indicator value.

Leaves and nuts, about half actual size

Native poplar   Desert poplar, Mustard tree

Codonocarpus cotinifolius

Family: Gyrostemonaceae

GENERAL DESCRIPTION: Native poplar is a tree which grows to about 8 m. The bark is smooth, often greyish green, but sometimes cream-coloured. It strongly resembles the European poplar, and tends to be conical in outline. It is short-lived, growing rapidly to its full height, and may lean markedly to one side as it develops. The leaves are green to greyish green, always broad and somewhat rounded, but occasionally slightly pointed. They are up to 5 cm long and are borne in an alternate fashion on long stalks. The native poplar has separate male and female plants. The male flowers are borne on

Native poplar, about 4 m high

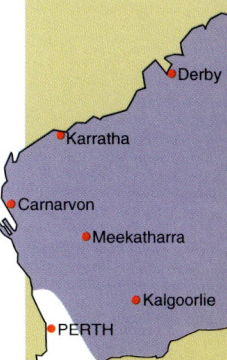

short stalks which arise in the axils of the stems and branches. The female flowers are very prominent and occur in open clusters, particularly near the apex of the tree. The fruits are bell-shaped, distinctively segmented, about 1 cm across and about 1.5 cm long. One seed is found in each of the segments and these eventually fall free from each other to the ground.

HABITAT: Native poplar occupies two distinct habitats. It occurs on limestone or 'opaline country' in major drainage tracts, where it is prominent above limestone wattle, pebble bush and sago bush. It is also an important plant on spinifex sandplains, where it appears in large numbers after fire.

FORAGE VALUE: Native poplar does not have any forage value since it is rarely eaten. It has been suspected of causing stock deaths, but tests for poisonous principles have proved negative.

INDICATOR VALUE: Native poplar has no indicator value.

Almost mature fruit, about actual size

POVERTY AND FUSCHIA BUSH GROUP

Berrigan   Long-leaved eremophila

Eremophila longifolia

Family: Myoporaceae

GENERAL DESCRIPTION: Berrigan is a tall shrub or small tree, which grows to 6 m. The dull grey-green leaves have a drooping habit and a sharp tip. They can be up to 15 cm long, but are rarely more than 1 cm wide. The tubular flowers are reddish brown and are about 2 cm long, with the petal tips curling backwards. The style, but not the stamens, is extended from the mouth of the flower. The fruit is oval, about 8 mm in diameter and contains a hard, woody seed.

Berrigan, about 4 m high

HABITAT: Berrigan is found in a variety of habitats in limestone country and in creeklines on hard mulga country. It is found on both calcareous and acidic loams.

FORAGE VALUE: It is readily eaten by stock, but is not common and therefore not an important dietary component of the pasture. It is very drought tolerant and its foliage persists during dry conditions. It may contain up to 14 per cent crude protein.

INDICATOR VALUE: Because it is uncommon, it is not a sensitive indicator of range condition. Under heavy grazing, populations decline.

A flower, about actual size

Almost mature fruit, about actual size

POVERTY AND FUSCHIA BUSH GROUP

Pixie bush

Eremophila oldfieldii

Family: Myoporaceae

GENERAL DESCRIPTION: Pixie bush is an erect, tall shrub or small tree, growing to 4 m. The dark green leaves, which are held upright, are 10 cm long and 4 mm wide. They are flattened and come to a point. A 2 cm long, s-shaped stalk holds each flower. The tubular flowers are brick red in the upper portion, tending to yellow towards the base. They end in five lobes, four above and the fifth curled back below. The style and stamens extend about 1 cm out from the mouth of the flower. The seeds are held in a hard fruit, 5 mm in diameter.

Pixie bush, about 2 m high

HABITAT: Pixie bush grows on breakaways, limestone country and on plains with shallow soils over granite. It grows with cotton bush, mulga and a wide range of non-saline shrubs.

FORAGE VALUE: It is attractive to stock and a useful feed. It is drought resistant, but much of the foliage is out of reach of stock.

INDICATOR VALUE: It has no indicator value.

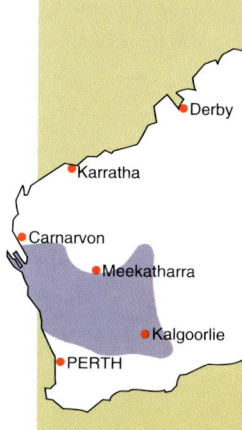

A flower, about actual size

Gum trees

Eucalyptus camaldulensis River red gum
Eucalyptus salmonophloia Salmon gum
Eucalyptus salubris Gimlet
Eucalyptus striaticalyx Cue york gum

Family: Myrtaceae

GENERAL NOTES: Gum trees dominate the landscape in the southern Goldfields, but in the remainder of the area their distribution is restricted to specialised habitats. They have been heavily cut for building timber and fuel in the past. Very few are suitable for fence posts, an exception being the black heart (*E. coolabah*), of the Pilbara. Most of the eucalypt forest in the Kalgoorlie, Coolgardie and Norseman district is cut over forest which has coppiced.

REHABILITATION: Gums are not recommended for rehabilitation programs unless they are planted in areas receiving run-on water, such as at the base of vertical drainage systems.

E. camaldulensis, River red gum: A tree up to 15 m, growing along watercourses where the water-table is shallow.

E. salmonophloia, Salmon gum: A tree up to 20 m high; it is restricted to the wheatbelt and the southern Goldfields. It forms an attractive, open woodland over an under-storey of bladder saltbush.

River red gum, about 15 m high

Salmon gum, about 15 m high

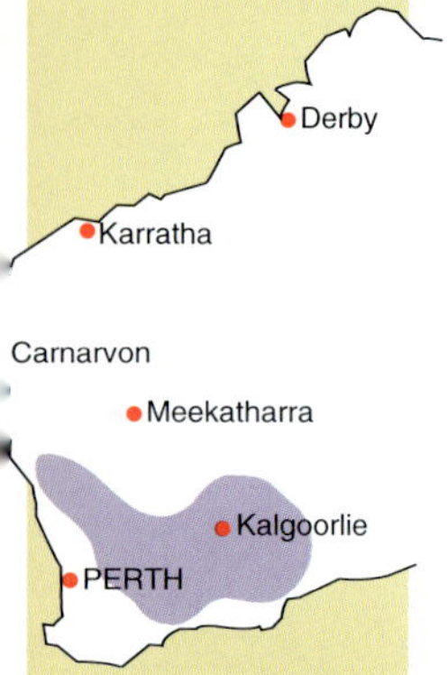

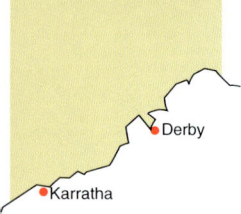

E. salubris, Gimlet: A shrubby tree with two to four trunks growing up to 10 m; it has fluted stems and attractive smooth, reddish, olive green bark. It grows with saltbushes and bluebush.

Gimlet, about 8 m high

E. striaticalyx, Cue york gum: A tree growing to 15 m, with one to three main stems and rough, black bark on main branches and reddish, smooth bark above. It is confined to strongly alkaline or gypsum-rich soils.

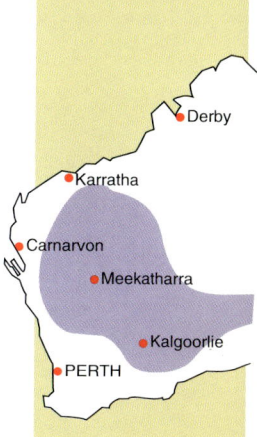

Cue york gum, about 10 m high

Christmas tree Waterbush

Grevillea berryana

Family: Proteaceae

GENERAL DESCRIPTION: Christmas tree grows to 4 m and up to 8 m wide. It has a rough, grey bark and masses of white flowers. The leaves are alternate along the stem and consist of either one long, thin leaf or a multi-branched leaf 3–15 cm long, but only 3 mm wide. The flowers are produced at the end of stems in large, white, cylindrical spikes from October to December, depending on location. The fruit is a dark brown, very woody, rounded pod. It contains two seeds, each of which is surrounded by a membranous wing.

Christmas tree, about 4 m high

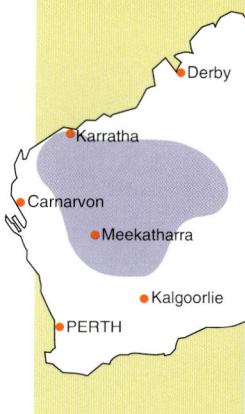

HABITAT: Christmas tree is restricted to sandplains that receive run-on water. It is associated with umbrella mulga, Wilcox bush and wanderrie grasses.

FORAGE VALUE: It is not known to be edible to stock.

INDICATOR VALUE: It has no indicator value.

NOTE: It is known as Christmas tree because it generally flowers at Christmas.

Leaves and persistent fruits, about one-quarter actual size

Sand-dune grevillea

Grevillea stenobotrya

Family: Proteaceae

GENERAL DESCRIPTION: Sand-dune grevillea is a low, upright tree which grows to 6 m and is generally restricted to sandplains and associated dunes. The erect leaves are 5–25 cm long and up to 2 mm wide. Cream flowers, which are 3–4 mm long, are held in groups of upright terminal spikes. They occur in September and October. The brittle, woody pods are 1–1.5 cm long and have a short, recurved beak. Two winged seeds are held in each pod. The pods break open when dry and the seeds are borne away by the wind.

Sand-dune grevillea, about 5 m high

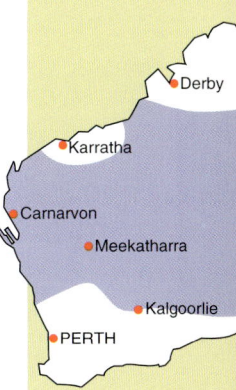

HABITAT: Sand-dune grevillea grows on sandplains and on sand dunes with spinifex in the Pilbara and with pine and spinifex wattle in the south.

FORAGE VALUE: It is not known to be eaten by stock.

INDICATOR VALUE: It has no indicator value.

Leaves, flowers and fruits, about one-quarter actual size

Beefwood

Grevillea striata

Family: Proteaceae

GENERAL DESCRIPTION: Beefwood is a tree which grows to 15 m under favourable conditions. It has a dark grey, deeply fissured bark and the base in large specimens may have a diameter of more than 60 cm. It usually has a single trunk and produces an open crown of drooping, flat, leathery, dull green leaves, up to 25 cm long and 5 mm wide. Numerous flowers are borne in 10 cm long creamy panicles. These produce small, woody, pod-like fruits. Beefwood also occurs in the Kimberley as a low tree, which rarely exceeds 3–4 m. It gets its common name from the colour of the freshly cut heartwood, which resembles raw beef.

Beefwood, about 10 m high

HABITAT: Beefwood is found in drainage depressions and watercourses in most mulga communities.

FORAGE VALUE: Since most leaf material is out of the grazing range of stock, beefwood has limited forage value.

INDICATOR VALUE: Beefwood has no indicator value. It appears to be a very long-lived tree and survives even the most extreme droughts.

Open pods, about half actual size

Needlebush Standback

Hakea preissii

Family: Proteaceae

GENERAL DESCRIPTION: Needlebush is a tall shrub or small tree, which can be 6 m high, but more commonly is smaller. The leaves are bluish green, round in cross-section, 2 mm in diameter and 1–3 cm long. They terminate in a sharp spine which is about 3 mm long. The flowers are yellow and are produced in masses of 20 or more flowers at the junction of the leaves and stems. The fruits, sometimes said to resemble the body of a snail, are very woody, up to 3 cm long and have two horn-like projections on the tip. The fruits split to release two flat seeds with papery wings.

Needlebush, about 4 m high

HABITAT: Needlebush is found on sand hills around salt lakes on saline stony plains and on floodplains and valley floors with clays, texture contrast soils and loams. It grows with Gascoyne bluebush, saltbushes, sago bush, curara and prickly acacia.

FORAGE VALUE: Needlebush has no forage value.

INDICATOR VALUE: It is common on alkaline soils that receive run-on water, but increases aggressively, forming dense thickets, when more palatable associated plants are overgrazed and removed. Where needlebush is a component of pastures, great care must be taken to ensure that overgrazing, which would encourage it, does not take place. In the Gascoyne it has already overrun much valuable pastoral land formerly covered with bluebush and saltbush. When found in large numbers, needlebush is a reliable indicator of poor range condition.

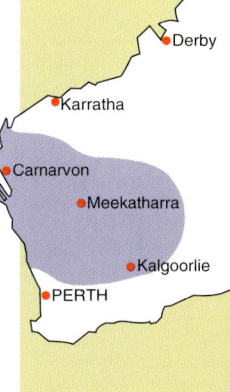

Unopened and opened fruit, about two-thirds actual size

Corkwood

Hakea suberea

Family: Proteaceae

GENERAL DESCRIPTION: Corkwood is a tree of gnarled appearance, which grows to 8 m. It has a characteristically deeply fissured, corky bark, often up to 3 cm thick. The leaves are 20–50 cm long, round in cross-section and 2–3 mm in diameter. They tend to be pendulous and generally lack a point at the tip. The flowers are yellowy green panicles, which arise in the axils of the stems and the leaves. They are up to 15 cm long and consist of many individual flowers held on a central stem. Each woody fruit is about 3.5 cm long and 2.5 cm wide, slightly ovoid and curved towards the end.

Corkwood, about 6 m high

It splits open at maturity to reveal two membranous winged seeds embedded one in each half.

HABITAT: Corkwood is found in floodplains and watercourses and on the sides of some stony rises and hills.

FORAGE VALUE: Corkwood has no forage value, as it is generally out of reach of stock.

INDICATOR VALUE: Corkwood has no indicator value.

Opened fruits, about half actual size

Trunk, showing cork-like bark

Mingah bush Rosewood, Bullock bush

Heterodendrum oleaefolium

Family: Sapindaceae

GENERAL DESCRIPTION: Mingah bush is a shrub or small tree, which grows to about 5 m but is usually less. The brown or grey bark is deeply fissured. The leaves are somewhat thick, stiff, lance-shaped and oblong, and about 10 cm long and up to 1 cm wide. They are a greyish green and have minute hairs on both sides which are difficult to see with the naked eye. The small, yellowish white flowers cluster at the ends of the branches and have no petals. The shiny red fruits consist of up to four globular lobes, each containing one black, glossy seed.

Mingah bush, about 3 m high

HABITAT: Mingah bush is usually found on alkaline soils and occurs with bluebushes and currant bushes in saltbush communities. It is also found in some mulga communities on deeper soils. It usually harbours large ant populations, hence its common name 'mingah', from an Aboriginal language meaning 'ant'.

FORAGE VALUE: It is a first-class fodder tree and is always heavily grazed. It is also known to be cyanogenic and can cause deaths from cyanide poisoning. Sheep appear to be more susceptible than cattle; horses are not affected. The toxic effects are only seen when stock are hungry and have little else to eat. Because it is always kept heavily grazed, it is unlikely to cause poisoning of stock in the paddock.

INDICATOR VALUE: Mature plants are long-lived and have no indicator value. The presence of young plants indicates improving range condition.

Leaves and flowers, about actual size

Native willow   Snotty gobbles, Cheesewood

Pittosporum phylliraeoides

Family: Pittosporaceae

GENERAL DESCRIPTION: Native willow is a shrub or small tree, which can grow to 6 m. It has gracefully drooping branches reminiscent of the garden or cultivated willow and a whitish or mottled trunk. The leaves are narrow, flat, a shiny green colour and are 5–10 cm long and up to 1 cm wide. They end in a short, hooked point. The cream or pale yellow flowers are borne singly or in clusters and are about 1 cm long. They are tubular in the bottom half, but have definite petals in the upper portion. The fruits hang pendulously from the branches and consist of yellow-orange capsules which dry and split open at maturity, revealing red seeds.

Native willow, about 5 m high

Its habit makes it prominent in the shrubland environment. It may be confused with spreading wattle at a distance since this acacia can also adopt the graceful willow-like habit.

HABITAT: Native willow is found in most communities, with the exception of the spinifex sandplains, but it is rarely common. It grows on a wide range of soil types from loams to clays, including calcrete platforms. It reaches its maximum height along the banks of watercourses and at the base of some granite tors and beneath breakaways.

FORAGE VALUE: It is sought after by cattle, but is unlikely to be grazed by sheep since the leaves are usually out of reach. It has a crude protein content of 8–11 per cent and a digestibility of 60 per cent.

INDICATOR VALUE: Native willow has no indicator value.

Fruit, about half actual size

Sweet quandong

Santalum acuminatum

Family: Santalaceae

GENERAL DESCRIPTION: Sweet quandong is a small tree or tall shrub, which resembles the true sandalwood, but the wood does not have the strong aromatic smell of true sandalwood. It grows to about 5 m, and has an open crown and pendulous branches. The bark is mid-grey and hardly fissured. The greyish green leaves are thickish and are arranged in pairs opposite each other. They are broadly oval, up to 6 cm long and 3 cm wide. The small flowers are borne in short panicles at the ends of the branches. They produce rounded, dark purplish, plum-like fruits about 1 cm in diameter. Each has a large, rough, globular stone. The plant is parasitic on other vegetation, and in

Sweet quandong, about 4 m high

common with other parasitic trees, the leaves are cool on a hot day in contrast to the leaves of surrounding non-parasitic bushes and trees. It is frequently subjected to severe insect attack.

HABITAT: It occurs in mulga communities on rocky hills and on plains, but it is never common.

FORAGE VALUE: The leaves are relished by stock and most sweet quandongs are heavily grazed.

INDICATOR VALUE: It has no known indicator value.

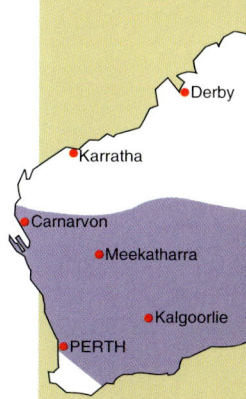

Leaves and fruits, about one-quarter actual size

Sandalwood

Santalum spicatum

Family: Santalaceae

GENERAL DESCRIPTION: Sandalwood is a tree up to 4 m high, with rough, almost black bark and, in sheep country, a round umbrella of foliage. The leaves are olive green, spear-shaped, 2–7 cm long and 3–15 mm wide and relatively thick and greasy to the touch. They are brittle, easily breaking between the thumb and forefinger. The flowers are white, very small and are carried in groups at the ends of branchlets. These mature into round, red-stoned, plum-like fruits, 1.5–2 cm in diameter, which have a 3 mm thick

Sandalwood, about 3 m high

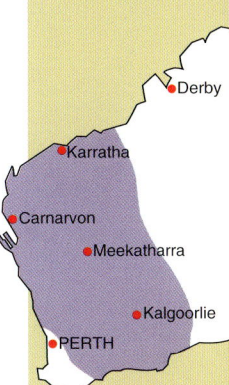

flesh over a hard, smooth stone. Sandalwood is parasitic on the roots of other plants. On the hottest days the leaves of sandalwood will be cool to the touch, since the plant does not expend energy on mineral uptake from the soil.

HABITAT: Sandalwood is found in a wide variety of habitats, but it prefers better-watered environments in woodlands, and it avoids rocky slopes and hills.

FORAGE VALUE: It is eagerly grazed by sheep, cattle, goats and rabbits, and its foliage persists well under dry conditions. It contains about 17 per cent crude protein.

INDICATOR VALUE: It never occurs in sufficient quantities to be an important indicator species. Grazing and sandalwood survival do not appear to be compatible, since all seedling trees are selectively removed by stock.

NOTE: It is the sandalwood of commerce and large quantities are sold overseas each year.

Mature, dried out fruit, about actual size

The origin of the generic names

The order of the plants in each of the sections of this book has, except for the everlastings, been determined by their Latin scientific names. The system of using two names to describe each plant — first the genus, then the species — was introduced by Carl Linnaeus.

In the systematic classification of plants that followed Linnaeus's work of 1753, botanists named plants to honour people (*Grevillea* for Charles Greville, a vice-president of the Royal Society of London), to call attention to some outstanding characteristic of the group (*Ptilotus*, from the Greek meaning winged or feathered), or to preserve an old Greek, Latin or Arabic name (*Acacia*, a prickly species growing in Egypt).

The following list describes the origin of the generic names used in this book. The translations come from Black's *Flora of South Australia*, 1948, and from Sharr, F.A., *Western Australian Plant Names and their Meanings*, 1988.

Name	Meaning
Acacia	from Greek, name of some prickly species growing in Egypt
Amphipogon	from Greek, two parts of the flowering spikelet have hairs (*amphi* both, *pogon* beard)
Angianthus	from the Greek, *aggeion* a vessel, and *anthos* a flower; referring to the hairs around the seed of one species which are grouped in a cup
Aristida	from Latin, *arista* an awn; drawing attention to the trifid (split into three parts) awn
Asphodelus	from Greek, *asphodelos* a lily-like plant growing in the Grecian place of after-life
Atriplex	from Latin, meaning spinach or the garden orach
Brachychiton	from Greek, *brachys* short, *chiton* an undergarment; referring to the loose outer covering of the seed
Brachycome	from Greek, *brachys* short, *come* hair; referring to the short hairs around the seed
Calandrinia	after Jean-Luis Calandrini, professor of botany and mathematics at Geneva
Callitris	from Greek, *kallos* beautiful, *treis* three; the leaves are arranged symmetrically in threes
Calytrix	from Greek, *kalyx* cup, *thrix* a hair; the sepals end in a long hair or bristle
Canthium	from Canti, the name of one species in Malabar

Cassia	from Graeco-Latin, from the Hebrew, *qase'ah* a type of cinnamon bark, and a fragrant shrub
Casuarina	from Latin *Casuarius* and Malay *casuari* a cassowary; likening the drooping branches to the drooping feathers of this bird
Cenchrus	from Greek, *cegchros* a type of millet, cenchrus is in the same group
Cephalipterum	from Greek, *cephale* head, *pteron* wing; refers to the winged nature of the compound head of flowers
Chenopodium	from Greek, *chen* a goose, *pous* a foot; the shape of the leaves of some species
Chrysopogon	from Greek, *chrysos* gold, *pogon* beard; refers to the tuft of hairs at the base of the flower head
Codonocarpus	from Greek, *codon* bell, *carpos* fruit; refers to the bell-like fruits
Corchorus	from Greek, *coreo* to purge; one species has laxative qualities
Cratystylis	from Greek, *cratys* strong, *stylos* pillar; referring to the strong style of the flower
Cymbopogon	from Greek, *cymbe* boat, *pogon* beard; refers to boat-shaped bract enclosing the seed head
Dactyloctenium	from Greek, *dactylos* a finger, *ktenion* a little comb; refers to finger-like and comb-like flowering heads
Danthonia	after Etienne Danthoine, a botanist of Marseilles
Disphyma	from Greek, *dissos* double, *phyma* tumour; refers to two wart-like bodies at the apex of the ovary
Dissocarpus	from Greek, *dissos* double, *carpos* fruit; referring to the paired fruits
Dodonaea	from Greek, *dysphanes* obscure; in reference to the small flowers
Enchylaena	from Greek, *enkhylos* succulent, *laina* a cloak; referring to the succulent fruit
Enneapogon	from Greek, *ennea* nine, *pogon* a beard; refers to the nine bearded awns on the seed
Eragrostis	from Greek, *eros* love, *agrostis* grass; referring to the elegant formation of the spikelets
Eremophila	from Greek, *eremos* desert, *phileo* to love; loving desert places where most species are found
Eriochiton	from Greek, *erion* wool, *chiton* a garment; referring to the woolly perianth or seed covering
Erodium	from Greek, *erodios* the heron; referring to the heron-like shape of the seed
Eucalyptus	from Greek, *eu* well, *calyptos* covered; referring to the cap or lid which covers the stamens in the bud

Eulalia	from Greek, *eu* well, *lalia* appearance; a handsome grass
Euphorbia	after the Greek physician Euphorbos, who discovered medicinal uses of some species in Africa
Exocarpos	from Greek, *exo* outside of, *carpos* fruit; referring to the fleshy stalk of the flower
Frankenia	after Johan Frankenius, professor of botany at Upsala, Sweden, in the 17th century
Gastrolobium	from Greek, *gastros* paunch, *lobos* a pod; referring to the swollen, ovoid pod
Grevillea	after Charles Greville, vice-president, Royal Society of London
Gunniopsis	after the genus *Gunnia* and Greek, *opsis* meaning like; name given when the genus was revised
Hakea	after Christian Ludwig von Hake, German patron of botany
Halosarcia	from Greek, *halos* salt, *sarx* flesh; referring to the succulent stems which contain much salt
Heterodendron	from Greek, *heteros* different, *dendron* tree; referring to the variable nature of the genus
Hyalosperma	from Greek, *hyalos* shining, *sporos* seed; referring to the shining seed
Ipomoea	from Greek, *ipos* a type of worm, *homoios* resembling
Lawrencia	after Robert William Lawrence, a botanist from Tasmania
Lycium	from Greek, *lycion* the buckthorn
Maireana	after the French naturalist Charles Lemaire
Marsdenia	after William Marsden, who collected specimens in Sumatra and Bencoolen; secretary of the British Admiralty
Marsilea	after L.F. Marsigli, an Italian naturalist, died 1730
Monachather	from Greek, *monos* single
Muehlenbeckia	after Gustav Muehlenbeck, an Alsatian botanist
Nitraria	from Latin, *nitrum* saltpetre; after the salt flats in Siberia where the plant was first found
Olearia	from Latin, *olea* olive tree; the leaves of the first species described resembled an olive
Pimelea	from Greek, *pimele* fat; probably referring to the fleshy seed leaves
Pittosporum	from Greek, *pitta* pitch, *sporos* seed; referring to the sticky pulp surrounding the seed
Plectrachne	from Greek, *plectron* a cock's spur, *achne* chaff; alluding to the stiff, three-awned flowering glume in the spikelet
Podolepis	from Greek, *podos* foot, *lepis* scale; the outside whorl of bracts of the flower have a stalk

Ptilotus	from Greek, *ptilotos* winged; meaning soft winged, presumably of the highly coloured flowers
Rhagodia	from Greek, *rhagodes* bearing berries; after the bunches in which the fruits are held
Rhodanthe	from Greek, *rhodos* red, *anthos* flower; red flowers
Salsola	from Latin, *salsus* salted; some species grow in salty ground
Santalum	from Arabic, *sandal*; the Indian sandalwood
Sarcostemma	from Greek, *sarx* flesh, *stemma* a headdress; referring to the fleshy outgrowths from the petals of the flower
Scaevola	from Latin, commemorating C. Mucius Scaevola (507 BC), who burned off his right hand; the name refers to the one-sided appearance of the flower
Schoenia	after Johannes Schoen, an eye specialist of Hamburg, early 19th century
Sclerolaena	from Greek, *scleros* hard, *chlaena* a covering; referring to the hard, woody fruits
Solanum	from Latin, *solanum* the deadly nightshade which is a solanum
Spartothamnella	like the genus *Spartothamnus*, the branches of which were used for rope making
Sporobolus	from Greek, *sporos* seed, *bolos* throwing; as the seed is easily shed
Stipa	from Greek, *stype* tow; referring to the tow or hemp obtained from Esparto grass, *S. tenacissima*
Stylobasium	from Greek, *stylos* a pillar, *basis* the ground; the style rises from the base of the ovary and not from the top
Swainsona	after Isaac Swainson, who maintained a private botanical garden near Twickenham, UK, about the year 1789; *Themeda* the Arabic name for this grass
Trianthema	from Greek, *treis* three, *anthimon* flower; the flowers sometimes grow in threes
Tribulus	from Latin, *tribulus* the four-pointed military instrument called the 'caltrop' employed to lame advancing cavalry as one of its four points always projected upwards when thrown on the ground
Triodia	from Greek, *treis* three, *odous* tooth; referring to the three teeth of the flowering glume in the spikelet
Velleia	after Thomas Velley, an English botanist
Waitzia	after Karl Friedrich Waitz, privy councillor of the Duchy of Saxe-Altenburg, Germany
Zygophyllum	from Greek, *zygon* the yoke or crossbar used with oxen, *phyllon* a leaf; each leaf has a pair of leaflets

INDEX
OF SCIENTIFIC NAMES AND AUTHORITIES

A

Acacia

acuminata subsp. *burkittii*	Benth.	368
anastema	Maslin	370
aneura	F. Muell. ex Benth.	367, 372
brachystachya	Benth.	376
citrinoviridis	Tind. & Maslin	378
coolgardiensis	Maiden	380
coriacea	DC.	382
craspedocarpa	F. Muell.	384
cuspidifolia	Maslin	386
cyperophylla	F. Muell. ex Benth.	388
eremaea	C.R.P. Andrews	419
grasbyi	Maiden	390
hemiteles	Benth.	392
inaequilatera	Domin.	394
intorta	Maslin	419
kempeana	F. Muell.	396
ligulata	Cunn. ex Benth.	398
linophylla	W. Fitzg.	367, 400
murrayana	F. Muell. ex Benth.	402
papyrocarpa	Benth.	404
pruinocarpa	Tind.	406
quadrimarginea	F. Muell.	408
ramulosa	W. Fitzg.	401
sclerosperma	F. Muell.	410
sibilans	Maslin	404
subtesserogona	Tind. & Maslin	412
tetragonophylla	F. Muell.	414
translucens	Cunn. ex Hook.	156, 367

victoriae	Benth.	416
xiphophylla	E. Pritzel.	367, 418
Amphipogon caricinus	F. Muell.	100
Angianthus milnei	Benth.	94
Aristida contorta	F. Muell.	102
Atriplex		
acutibractea	R. Anderson	26
amnicola	Paul G. Wilson	158
bunburyana	F. Muell.	160
codonocarpa	Paul G. Wilson	28
holocarpa	F. Muell.	30
nummularia	Lindl.	162
semilunaris	Aellen	32
stipitata	Benth.	164
vesicaria	Heward	166

B

Brachychiton gregorii	F. Muell.	420
Brachycome ciliocarpa	W. Fitzg.	94

C

Calandrinia polyandra	Benth.	34
Callitris glaucophylla	F. Muell.	422
Canthium		
attenuatum	R. Br. ex Benth.	425
latifolium	F. Muell. ex Benth.	424
lineare	E. Pritzel.	425
Carrichtera annua	(L.) DC.	36
Cassia		
artemisioides	Gaudich in DC.	170
chatelainiana	Gaudich	172
desolata	F. Muell.	174

hamersleyensis	D.E. Symon	176
helmsii	D.E. Symon	169, 178
luerssenii	Domin.	169, 180
nemophila	Cunn. ex Vogel	169, 182
oligophylla	F. Muell.	169, 184
sturtii	R. Br.	186
Casuarina obesa	Miq. in Lehm.	426
Cenchrus		
ciliaris	L.	104
setiger	M. Vahl	105
Cephalipterum drummondii	A. Gray	94
Chenopodium		
auricomum	Lindl.	188
gaudichaudianum	(Moq.) Paul G. Wilson	190
Chrysopogon fallax	S.T. Blake	106
Codonocarpus cotinifolius	(Desf.) F. Muell.	428
Commicarpus australis	Meikle	38
Corchorus walcottii	F. Muell.	192
Cratystylis		
conocephala	(F. Muell.) S. Moore	194
subspinescens	(F. Muell. et Tate) S. Moore	196
Cymbopogon ambiguus	A. Camus	108

D

Danthonia caespitosa	Gaudich in Freyc.	110
Disphyma crassifolium		
subsp. *clavellatum*	(Haw.) Chinn.	40
Dissocarpus paradoxus	(R. Br.) F. Muell.	42
Dodonaea		
lobulata	F. Muell.	198
viscosa	Jacq.	200
Dysphania		
kalpari	Paul. G. Wilson	44
rhadinostachya	(F. Muell.) A.J. Scott	44

E

Enchylaena tomentosa	R. Br.	202
Enneapogon caerulescens	(Gaudich) N. Burb.	112
Enteropogon acicularis	B. Simon	114
Eragrostis		
dielsii	Pilger ex Diels & E. Pritzel	116
eriopoda	Benth.	118
lanipes	C.E. Hubb.	120
setifolia	Nees	122
xerophila	Domin.	124
Eremophila		
arachnoides	Chinn.	205, 249
clarkei	A.F. Oldfield	221
compacta	S. Moore	206
'*crenulata*'	ms.	208
cuneifolia	Kraenzlin	210
delisseri	F. Muell.	212
eriocalyx	F. Muell.	231
'*flaccida*'	ms.	217
forrestii	F. Muell.	214
fraseri	F. Muell.	216
freelingii	F. Muell.	218
georgei	Diels in Diels & E. Pritzel	220
glabra	(R. Br.) Ostenf.	222
'*glandulifera*'	ms.	224
granitica	S. Moore	226
laanii	F. Muell.	228
latrobei	F. Muell.	230
linearis	Chinn.	205, 232
longifolia	(R. Br.) F. Muell.	430
macmillaniana	C. Gardner	234
maculata	(Ker Gawler) F. Muell.	236
maitlandii	F. Muell. ex Benth.	238
margarethae	S. Moore	240

miniata	C. Gardner	242
oldfieldii	F. Muell.	432
pantonii	F. Muell.	249
platycalyx	F. Muell.	244
pterocarpa	W. Fitzg.	246
ramiflora	B. Dell	217
scoparia	(R. Br.) F. Muell.	248
spectabilis	C. Gardner	250
Eriachne		
aristidea	F. Muell.	126
benthamii	(Domin.) Hartley	128
flaccida	Hartley	130
helmsii	Hartley	132
Eriochiton sclerolaenoides	(F. Muell.) F. Muell. ex A.J. Scott	46
Erodium crinitum	Carolin	48
Eucalyptus		
camaldulensis	Dehnh.	434
coolabah	Blakely	434
salmonophloia	F. Muell.	434
salubris	F. Muell.	434
striaticalyx	W. Fitzg.	434
Eulalia fulva	(R. Br.) Kuntze	134
Euphorbia drummondii	Boiss	50
Exocarpos aphyllus	R. Br.	252

F

Frankenia spp.		254

G

Gastrolobium		
brevipes	M.D. Crisp	257
laytonii	J. White	256

Grevillea			
berryana		Ewart & J. White	438
deflexa		F. Muell.	258
inconspicua		Diels in Diels & E. Pritzel	260
stenobotrya		F. Muell.	440
striata		R. Br.	442
Gunniopsis quadrifida		(F. Muell.) Pax in Engl. & Prantl.	262

H

Hakea			
preissii		Meissner	444
suberea		S. Moore	446
Halosarcia doleiformis		Paul. G. Wilson	264
Heterodendrum oleaefolium		Desf.	448

I

Ipomoea muelleri	Benth.	52

L

Lawrencia			
helmsii		(F. Muell. & Tate) N. Lander	266
squamata		Nees ex Miq.	268
Lycium australe		F. Muell.	270

M

Maireana			
amoena		(Diels) Paul G. Wilson	274
aff. *tomentosa*		Moq.	276
aphylla		(R. Br.) Paul G. Wilson	278
atkinsiana		(W. Fitzg.) Paul G. Wilson	280
brevifolia		(R. Br.) Paul G. Wilson	282
carnosa		(Moq.) Paul G. Wilson	54
convexa		Paul G. Wilson	284
georgei		(Diels) Paul G. Wilson	273, 286

glomerifolia	(F. Muell. & Tate) Paul G. Wilson	273, 288
lanosa	(Lindl.) Paul G. Wilson	290
melanocoma	(F. Muell.) Paul G. Wilson	292
planifolia	(F. Muell.) Paul G. Wilson	294
platycarpa	Paul G. Wilson	296
polypterygia	(Diels) Paul G. Wilson	273, 298
pyramidata	(Benth.) Paul G. Wilson	300
sedifolia	(F. Muell.) Paul G. Wilson	302
thesioides	(C. Gardner) Paul G. Wilson	304
tomentosa	Moq.	306
trichoptera	(J. Black) Paul G. Wilson	308
triptera	(Benth.) Paul G. Wilson	310
Marsdenia australis	(R. Br.) Druce	56
Marsilea drummondii	A. Braun	58
Monachather paradoxa	Steudel	136
Muehlenbeckia cunninghamii	(Meissner) F. Muell.	312

N

Nitraria billardierei	DC.	314

O

Olearia muelleri	Benth.	316

P

Pimelea microcephala	R. Br.	318
Pittosporum phylliraeoides	DC.	450
Plectrachne schinzii	Henrard	138
Podolepis canescens	Cunn. ex DC.	94
Ptilotus		
aervoides	(F. Muell.) F. Muell.	60
beardii	Benl.	320
divaricatus	(Gaudich) F. Muell.	322
exaltatus	Nees	62
lazaridis	Benl.	324

macrocephalus	(R. Br.) Poiret	64
obovatus	(Gaudich) F. Muell.	326
polakii	F. Muell.	328
polystachyus	(Gaudich) F. Muell.	66
rotundifolius	(F. Muell.) F. Muell.	330
schwartzii	F. Muell. ex Tate	332

R

Rhagodia
drummondii	Moq. in DC.	334
eremaea	Paul G. Wilson	336

Rhodanthe chlorocephala
subsp. *splendida*	(Hemsley) Wilson	94
floribunda	(Hemsley) Wilson	94

S

Salsola kali	L.	68

Santalum
acuminatum	(R. Br.) A. DC.	452
spicatum	(R. Br.) A. DC.	454

Sarcostemma viminale
subsp. *australe*	(R. Br.) P.I. Forst.	338

Scaevola
spinescens	R. Br.	340
tomentosa	Gaudich	342
Schoenia cassiniana	(Gaudich) Steetz. in Lehm.	94

Sclerolaena
bicornis	Lindl. in Mitch.	70
cuneata	Paul G. Wilson	72
densiflora	(W. Fitzg.) A.J. Scott	74
diacantha	(Nees) Benth.	76
eriacantha	(F. Muell.) Ulbr. in Engl.	78

eurotioides	(F. Muell.) A.J. Scott	80
hostilis	(Diels) Domin.	344
obliquicuspis	(R. Anderson) Ulbr. in Engl. & Prantl	82
patenticuspis	(R. Anderson) Ulbr. in Engl. & Prantl	84
Senecio magnificus	F. Muell.	346
Sida		
calyxhymenia	Gay ex DC.	348
corrugata	Lindl.	350
Solanum		
lasiophyllum	Dunal ex Poiret	352
orbiculatum	Dunal ex Poiret	354
Spartothamnella teucriiflora	(F. Muell.) Mold.	356
Sporobolus mitchellii	(Trin.) C.E. Hubb.	140
Stipa		
elegantissima	Labill.	142
scabra	Lindl.	144
Stylobasium spathulatum	Desf.	358
Swainsona		
formosa	(G. Don.) J. Thompson	86
pterostylis	(DC.) Bakh	88

T

Themeda triandra	Forsskal.	146
Thyridolepis multiculmis	(Pilger) S.T. Blake	148
Trianthema triquetra	Willd.	90
Tribulus platypterus	Benth.	360
Triodia		
basedowii	E. Pritzel	150
lanigera	Domin.	151
pungens	R. Br.	151, 152
wiseana	C. Gardner	151
intermedia	E. Cheel	151

V

Velleia rosea   S. Moore   92

W

Waitzia acuminata   Steetz in Lehm.   94

Z

Zygophyllum aurantiacum   (Lindl.) F. Muell.   362

NOTE: *Eremophila* 'crenuluta', *Eremophila* 'flaccida' and *Eremophila* 'glandulifera' are unpublished manuscript names provided by R.J. Chinnock.

INDEX
OF COMMON NAMES

A

Annual saltbush	32
Australian boxthorn	270
Australian sage	196

B

Ballart, leafless	252
Ball burr	42
Ball-leaf bluebush	273, 288
Balsam	50
Bandicoot grass	136
Banjine	318
Barbed wire grass	148
Bardi bush	416
Bead hopbush	198
Beefwood	442
Bell saltbush	28
Berrigan	430
Bindii, fluffy	80
Bindii, giant	344
Bindii, grey	76
Bindii, limestone	82
Bindii, silky	78
Bindii, tall	78
Bindii, woolly	46
Bindii, yellow	72
Birdwood grass	105
Black bluebush	300
Black gidgee	406
Black mulga	378
Bladder saltbush	166
Bloodbush	169, 184
Bluebush, ball-leaf	288
Bluebush, black	300
Bluebush, breakaway	276
Bluebush, brittle	274
Bluebush, bronze	280
Bluebush, cottony	54
Bluebush, downy	308
Bluebush, false	194
Bluebush, felty	306
Bluebush, five-pin	280
Bluebush, flat-leaved	294
Bluebush, Gascoyne	273, 298
Bluebush, George's	286
Bluebush, golden	273, 286
Bluebush, lax	304
Bluebush, low	294
Bluebush, mulga	284
Bluebush, northern	188
Bluebush, pearl	302
Bluebush, pink-seeded	308
Bluebush, pussy	292
Bluebush, short-leaf	282
Bluebush, shy	296
Bluebush, spiny	278
Bluebush, swamp	188
Bluebush, three-winged	310
Bluebush, woolly	290
Blunt-leaf cassia	178
Bohemia	386
Bottle washers	66
Bowgada	367, 400
Bowgada, false	376

471

Breakaway bluebush	276
Breelya	256
Brittle bluebush	274
Broad-leaf hopbush	200
Broad-leaved wanderrie grass	136
Bronze bluebush	280
Broom bush	248
Browntop, silky	134
Buck bush	68
Buck wanderrie grass	132
Buffel grass, WA	104
Bullock bush	448
Burr, goathead	70
Burra	216
Buttercups	184

C

Caltrop, corky bark	360
Cannon balls	42
Cassia, blunt-leaf	178
Cassia, creeping	176
Cassia, crinkled	169, 178
Cassia, desert	169, 182
Cassia, green	172
Cassia, grey	178
Cassia, silver	170
Cassia, straight leaf	174
Cassia, variable	186
Cassia, white	169, 180
Caustic bush	338
Caustic weed	50
Cheesewood	450
Christmas tree	438
Claypan grass	130
Climbing mulla mulla	322

Cogla	56
Common wallaby grass	110
Copperburr, spear-fruit	84
Corky bark caltrop	360
Corkscrew	48
Corkwood	446
Cotton bush	326
Cottony bluebush	54
Cottony saltbush	190
Couch, rats tail	140
Couch, river	140
Creekline miniritchie	388
Creeping cassia	176
Creeping wanderrie grass	120
Crinkled cassia	169, 178
Crowsfoot	48
Cue grevillea	260
Cue york gum	434
Curara	414
Curious bassia	42
Currant bush	340
Currant, wild	424

D

Daisy, Goldfields	316
Desert cassia	169, 182
Desert kurrajong	420
Desert poplar	428
Desert pride	231
Dillon bush	314
Downy bluebush	308
Dunna-dunna	266
Dwarf saltbush	28
Dwarf sida	350

E

Emu bush	236
Eremophila, long-leaved	430
Eremophila, warty-leaf	230
Eremophila, wedge-leaved	210
Eremophila, wing-fruited	246
Everlasting daisies	94

F

False bluebush	194
False bowgada	376
False wanderrie grass	126
Feather speargrass	142
Featherheads	64
Felt bush	214
Felted scaevola	342
Felty bluebush	306
Felty fuchsia bush	206
Fine leaf jam	368
Fine-toothed poverty bush	220
Fire wattle	402
Fish poison	360
Five-pin bluebush	280
Flannel bush	352
Flat-leaved bluebush	294
Flat-topped saltbush	28
Fluffy bindii	80
Four-o'clock	34
Frankenia	254
Frontage poverty bush	212
Fuchsia bush	236
Fuchsia bush, felty	206
Fuchsia bush, harlequin	232
Fuchsia bush, rock	218
Fuchsia bush, saline	205, 232
Fuchsia bush, warty	230

G

Gascoyne bluebush	273, 298
Gascoyne mulla mulla	328
George's bluebush	286
Giant bindii	344
Gidgee	370, 406
Gidgee, black	406
Gidgee, sandridge	370
Gidgee, spreading	412
Gimlet	434
Goathead burr	70
Golden beard grass	106
Golden bluebush	273, 286
Goldfields daisy	316
Granite poverty bush	244
Granite wattle	396, 408
Grapevine bush	342
Grass, bandicoot	136
Grass, broad-leaved wanderrie	136
Grass, buck wanderrie	132
Grass, buffel WA	104
Grass, claypan	130
Grass, creeping wanderrie	120
Grass, false wanderrie	126
Grass, golden beard	106
Grass, kangaroo	146
Grass, kerosene	102
Grass, lemon-scented	108
Grass, limestone	112
Grass, love	116
Grass, mulga	148
Grass, Murchison red	116
Grass, neverfail	122
Grass, red	116
Grass, ribbon	106
Grass, Roebourne Plains	124

Grass, silky browntop	134
Grass, soft wanderrie	148
Grass, spear	144
Grass, WA buffel	104
Grass, weeping	106
Grass, wind	102
Grass, windmill	114
Grass, wire wanderrie	118
Grass, woolly butt	118
Green cassia	172
Green crumbweed	44
Green mulla mulla	66
Grevillea, Cue	260
Grevillea, red	258
Grevillea, sand-dune	440
Grey bindii	76
Grey cassia	178
Grey copperburr	76
Grey corchorus	192
Grey fan leaf	268
Grey turpentine bush	234
Groundsel, showy	346
Gum trees	434

H

Hard spinifex	150
Hairy bindii	74
Harlequin fuchsia bush	232
Hop mulga	384
Hopbush, bead	198
Hopbush, broad-leaf	200
Hopbush, sticky	200
Hopbush, wild	198
Horse mulla mulla	332

J

Jam, fine leaf	368
Jam, raspberry	368
Jam, river	382

K

Kangaroo grass	146
Kanji bush	394
Kerosene grass	102
Kidney saltbush	164
Kite leaf	256
Knotty-butt neverfail	124
Kopi poverty bush	242
Kurara	414
Kurrajong, desert	420

L

Lake-fringe rhagodia	334
Lax bluebush	304
Leafless ballart	252
Lemon, wild	424
Lemon-scented grass	108
Lignum	312
Limestone bindii	82
Limestone grass	112
Limestone wattle	410
Long-leaved eremophila	430
Long greybeard grass	100
Love grass	116
Low bluebush	294
Low mulla mulla	320
Low rhagodia	334

M

Maroon bush	340
Mat mulla mulla	60

Mat spurge	50
Mingah bush	448
Miniritchie	390
Miniritchie, creekline	388
Morning glory, poison	52
Mulga	367, 372
Mulga, black	378
Mulga, hop	384
Mulga, red	388
Mulga, umbrella	376
Mulga bluebush	284
Mulga broom bush	356
Mulga grass	148
Mulla mulla, climbing	322
Mulla mulla, Gascoyne	328
Mulla mulla, green	66
Mulla mulla, horse	332
Mulla mulla, low	320
Mulla mulla, mat	60
Mulla mulla, narrow-leaved	322
Mulla mulla, purple	62
Mulla mulla, pussytail	64
Mulla mulla, royal	330
Mulla mulla, saline	324
Mulla mulla, tall	62
Murchison River poverty bush	228
Murchison red grass	116
Murray's wattle	402
Mustard tree	428

N

Naked lady	252
Nardoo	58
Narrow-leaved mulla mulla	322
Native pear	56
Native plum	424
Native poplar	428
Native willow	450
Needlebush	444
Neverfail	122
Nitre bush	314
Northern bluebush	188

O

Oat-eared spinifex	138
Old man saltbush	162

P

Parakeelya	34
Pearl bluebush	302
Pebble bush	358
Perennial tar vine	38
Pine	422
Pink poverty bush	224
Pink-seeded bluebush	308
Pinyuru	210
Pixie bush	432
Poison morning glory	52
Pop saltbush	30
Poplar, desert	428
Poplar, native	428
Poverty bush	156, 367
Poverty bush, fine-toothed	220
Poverty bush, frontage	212
Poverty bush, granite	244
Poverty bush, kopi	242
Poverty bush, Murchison River	228
Poverty bush, pink	224
Poverty bush, royal	210
Poverty bush, sandbank	240
Poverty bush, showy	250
Poverty bush, silver	246

Poverty bush, slender	230	Royal mulla mulla	330
Poverty bush, stony	218	Royal poverty bush	210
Poverty bush, tall	238	Ruby saltbush	202
Poverty bush, thin-leaved	226		
Poverty bush, waxy-leaf	208		
Prickly acacia	416		
Prostrate sida	350	**S**	
Punty bush	182	Sage	196
Purple mulla mulla	62	Sago bush	300
Purple vetch	88	Saline fuchsia bush	205, 232
Pussy bluebush	292	Saline mulla mulla	324
Pussy tail mulla mulla	64	Salmon gum	434
		Saltbush, annual	32
		Saltbush, bell	28
R		Saltbush, bladder	166
Ragged-leaf fan flower	342	Saltbush, cottony	190
Raspberry jam	368	Saltbush, dwarf	28
Rats tail couch	140	Saltbush, flat-topped	28
Red grass	116	Saltbush, kidney	164
Red grevillea	258	Saltbush, old man	162
Red mulga	388	Saltbush, pop	30
Red spinach	90	Saltbush, river	158
Rhagodia, lake-fringe	334	Saltbush, ruby	202
Rhagodia, low	334	Saltbush, scrambling	190
Ribbon grass	106	Saltbush, silver	160
River couch	140	Saltbush, swamp	158
River jam	382	Saltbush, tall	336
River red gum	434	Saltbush, thorny	336
River saltbush	158	Saltbush, toothed	26
Rock fuchsia bush	218	Samphire	264
Roebourne Plains grass	124	Samphire, sweet	262
Roly poly	68	Sandalwood	454
Rope plant	54	Sandbank poverty bush	240
Round-leaf pigface	40	Sand-dune grevillea	440
Round leaved solanum	354	Sandplain wattle	402
Rosewood	448	Sandridge gidgee	370

Scaevola, felted	342	Spreading gidgee	412
Scented oil grass	108	Spreading wattle	408
Scrambling saltbush	190	Standback	444
Sheoak	426	Stick-plant bush	356
Short-leaf bluebush	282	Sticky hopbush	200
Showy groundsel	346	Stony poverty bush	218
Showy poverty bush	250	Straight leaf cassia	174
Shrubby rice flower	318	Sturt pea	86
Shrubby twinleaf	362	Sturt's pigface	262
Shy bluebush	296	Sugar brother	380
Sida, dwarf	350	Swamp bluebush	188
Sida, prostrate	350	Swamp grass	128
Sida, tall	348	Swamp oak	426
Silky bindii	78	Swamp saltbush	158
Silky browntop	134	Sweet quandong	452
Silver bark wattle	410	Sweet samphire	262
Silver cassia	170		
Silver poverty bush	246		
Silver saltbush	160		
Silver speargrass	142	**T**	
Silvery emu bush	248	Tall bindii	78
Skeleton bark	360	Tall mulla mulla	62
Slender poverty bush	230	Tall poverty bush	238
Snakewood	367, 418	Tall saltbush	336
Snotty gobbles	450	Tall sida	348
Solanum, round leaved	354	Tan wattle	392
Soft wanderrie grass	148	Tar bush	222
Soft spinifex	138, 152	Thin-leaved poverty bush	226
Spear-fruit copperburr	84	Thorny saltbush	336
Speargrass	144	Three-winged bluebush	310
Speargrass, feather	142	Tomato, wild	354
Speargrass, silver	142	Toothed saltbush	26
Spinifex, hard	150	Travel bush	236
Spinifex, soft	138, 152	Turpentine bush	216
Spinifex wattle	380	Turpentine bush, grey	234
Spiny bluebush	278	Twinleaf, shrubby	362

U

Umbrella mulga	376
Umbrella wattle	398

V

Variable cassia	186
Velleia	92

W

WA buffel grass	104
Wait-a-while	386
Wanderrie grass, broad-leaved	136
Wanderrie grass, buck	132
Wanderrie grass, creeping	120
Wanderrie grass, false	126
Wanderrie grass, soft	148
Wanderrie grass, wire	118
Wanderrie wattle	396
Wanyu	400
Ward's weed	36
Warty fuchsia bush	230
Warty-leaf eremophila	230
Water bush (shrubby)	270
Water bush (tree)	438
Wattle, fire	402
Wattle, granite	396, 408
Wattle, limestone	410
Wattle, Murray's	402
Wattle, sandplain	402
Wattle, silver bark	410
Wattle, spinifex	380
Wattle, spreading	408
Wattle, tan	392
Wattle, umbrella	398
Wattle, wanderrie	396
Wattle, wiry	382
Waxy-leaf poverty bush	208
Wedge-leaved eremophila	210
Weeping grass	106
Western myall	404
White cassia	169, 180
White cypress pine	422
White top	110
Wilcox bush	214
Wild currant	424
Wild hopbush	198
Wild lemon	424
Wild tomato	354
Wind grass	102
Windmill grass	114
Wing-fruited eremophila	246
Wire wanderrie grass	118
Wiry wattle	382
Witchetty bush	396
Woolly bindii	46
Woolly bluebush	290
Woolly butt	118
Woolly-fruited copperburr	46

Y

Yellow bindii	72

DATE DUE

DUE DATE SUBJECT TO CHANGE
IF A RECALL IS REQUESTED

DEMCO, INC. 38-2931